282 Current Topics in Microbiology and Immunology

Editors

R.W. Compans, Atlanta/Georgia
M.D. Cooper, Birmingham/Alabama
H. Koprowski, Philadelphia/Pennsylvania
F. Melchers, Basel · M.B.A. Oldstone, La Jolla/California
S. Olsnes, Oslo · M. Potter, Bethesda/Maryland
P.K. Vogt, La Jolla/California · H. Wagner, Munich

Springer
*Berlin
Heidelberg
New York
Hong Kong
London
Milan
Paris
Tokyo*

H. Stenmark (Ed.)

Phosphoinositides in Subcellular Targeting and Enzyme Activation

With 20 Figures and 4 Tables

 Springer

Professor Harald Stenmark
Department of Biochemistry
The Norwegian Radium Hospital
Montebello
0310 Oslo
Norway
e-mail: stenmark@ulrik.uio.no

Cover Illustration by Gerald Hammond et al. (this volume):

Phosphoinositides and their soluble analogues inositol polyphosphates are widely distributed in the cell and function as essential co-factors for cytoplasmic and nuclear processes. The presence of a nuclear pool of PtdIns(4,5)P_2 (not associated with a membrane bilayer) is here demonstrated using a specific monoclonal antibody against this lipid; the staining reveals a speckled pattern resembling interchromatin granule clusters (IGC, in green) in the nuclei of HeLa cells. IGC represent a functional compartment of the nucleus (central scheme), together with nucleoli, Cajal bodies and their associated gemini of coiled bodies (see Hammond et al., this volume).

ISSN 0070-217X
ISBN 3-540-00950-7
Springer-Verlag Berlin Heidelberg New York

Library of Congress Catalog Card Number 72-152360

This work is subject to copyright. All rights are reserved, whether the whole or part of the material is concerned, specifically the rights of translation, reprinting, reuse of illustrations, recitation, broadcasting, reproduction on microfilm or in any other way, and storage in data banks. Duplication of this publication or parts thereof is permitted only under the provisions of the German Copyright Law of September 9, 1965, in its current version, and permission for use must always be obtained from Springer-Verlag. Violations are liable for prosecution under the German Copyright Law.

Springer-Verlag Berlin Heidelberg New York
a member of BertelsmannSpringer Science+Business Media GmbH

http://www.springer.de

© Springer-Verlag Berlin Heidelberg 2004
Library of Congress Catalog Card Number 15-12910
Printed in Germany. Not for sale

The use of general descriptive names, registered names, trademarks, etc. in this publication does not imply, even in the absence of a specific statement, that such names are exempt from the relevant protective laws and regulations and therefore free for general use.

Cover Design: Design & Production GmbH, Heidelberg
Typesetting: Stürtz AG, Würzburg
Production Editor: Angélique Gcouta, Berlin
Printed on acid-free paper   SPIN: 10922903   27/3020   5 4 3 2 1 0

Preface

Cells of the immune system are activated by a variety of stimuli that are derived from other cells, ingested material or from invading micro-organisms. A common denominator of such stimulation is that it causes a transient assembly of intracellular signalling complexes, which orchestrate cellular responses ranging from altered gene transcription to cell migration and phagocytosis. While some signalling complexes assemble in the cytosol or nucleoplasm, the majority of such assemblies occur on cellular membranes. Clearly, cells are equipped with machineries that enable a rapid and reversible recruitment of cytosolic proteins to specific intracellular membranes, such as the plasma membrane, phagosomes or endosomes. Besides protein–protein interactions, lipid–protein interactions are crucial in this context. In particular, phosphoinositides, which are phosphorylated derivatives of phosphatidylinositol, are important for recruiting and activating the right proteins at the right membranes. This volume *Current Topics in Microbiology and Immunology* focuses on the mechanisms of phosphoinositide-mediated protein recruitment to intracellular membranes. Recent advances in cell biology and bioinformatics have revealed the existence of several conserved protein modules, such as PH, FYVE, ENTH and PX domains, which endow proteins with the ability to bind specific phosphoinositides and thereby enables their targeting to specific membranes. It is fascinating to learn how this recruitment regulates receptor signalling, membrane trafficking, cytoskeletal function, chemotaxis and microbial killing—cellular functions that keep the immune system up and going. While the role of membrane-associated phosphoinositides in protein targeting has been well characterized, a new and unexpected role of phosphoinositides is also emerging. Several phosphoinositides have been detected in the nucleus, mainly outside the membrane bilayers. As discussed in the final review of this volume, evidence is accumulating that these nuclear phosphoinositides function as steric regulators of multi-enzymatic functions such as chromatin remodelling and pre-mRNA processing. This opens up a new and exciting area of research that promises to shed light on some of the most fundamental and complex processes in cell biology.

August 2003   Harald Stenmark

List of Contents

Phosphoinositide Involvement in Phagocytosis
and Phagosome Maturation
R.J. Botelho, C.C. Scott, and S. Grinstein 1

Regulation of Endocytosis by Phosphatidylinositol 4,5-Biphosphate
and ENTH Proteins
T. Itoh and T. Takenawa 31

Membrane Targeting by Pleckstrin Homology Domains
G.E. Cozier, J. Carlton, D. Bouyoucef, and P.J. Cullen 49

Protein Targeting to Endosomes and Phagosomes via FYVE
and PX Domains
H.C.G. Birkeland and H. Stenmark 89

Regulation of the Actin Cytoskeleton by $PI(4,5)P_2$ and $PI(3,4,5)P_3$
P. Hilpelä, M.K. Vartiainen, and P. Lappalainen 117

Roles of PI3K in Neutrophil Function
M.O. Hannigan, C.K. Huang, and D.Q. Wu 165

Nuclear Phosphoinositides and Their Functions
G. Hammond, C.L. Thomas, and G. Schiavo 177

Subject Index 207

List of Contributors

(Their addresses can be found at the beginning of their respective chapters.)

Birkeland, H.C.G.   89

Botelho, R.J.   1

Bouyoucef, D.   49

Charlton, J.   49

Cozier, G.E.   49

Cullen, P.J.   49

Grinstein, S.   1

Hammond, G.   177

Hannigan, M.O.   165

Hilpelä, P.   117

Huang, C.K.   165

Itoh, T.   31

Lappalainen, P.   117

Schiavo, G.   177

Scott, C.C.   1

Stenmark, H.   89

Takenawa, T.   31

Thomas, C.L.   177

Vartiainen, M.K.   117

Wu, D.Q.   165

Phosphoinositide Involvement in Phagocytosis and Phagosome Maturation

R. J. Botelho · C. C. Scott · S. Grinstein

Programme in Cell Biology, Hospital for Sick Children
and the Department of Biochemistry, University of Toronto,
Toronto, Ontario, M5G 1X8, Canada
E-mail: Sga@sickkids.ca

1	Introduction	2
2	Phagocytosis	2
3	Phagosome Maturation	4
4	Phosphoinositides	6
5	Phosphatidylinositol-4,5-bisphosphate	8
6	Phosphatidylinositol-3,4-bisphosphate and Phosphatidylinositol-3,4,5-trisphosphate	15
7	Phosphatidylinositol-3-phosphate	17
	References	21

Abstract Cells of the innate immune system engulf invading microorganisms into plasma membrane-derived vacuoles called phagosomes. Newly formed phagosomes gradually acquire microbicidal properties by a maturation process which involves sequential and coordinated rounds of fusion with endomembranes and concomitant fission. Some pathogens interfere with this maturation sequence and thereby evade killing by the immune cells, managing to survive intracellularly as parasites. Phosphoinositides seem to be intimately involved in the processes of phagosome formation and maturation, and initial observations suggest that the ability of some microorganisms to survive intracellularly is associated with alterations in phosphoinositide metabolism. This chapter presents a brief overview of phosphoinositides in cells of the immune system, their metabolism in the context of phagocytosis and phagosome maturation and their possible derangements during infectious pathogenesis.

1
Introduction

Phagocytosis is a dynamic and carefully orchestrated process whereby the plasma membrane encircles and internalizes particles ≥ 0.5 µm in diameter. The engulfed particles are sequestered within a unique intracellular organelle termed the phagosome. Although the basic process has been conserved during evolution, phagocytosis serves diverse functions in different organisms. It is employed by unicellular organisms, such as *Dictyostelium*, to feed on bacteria. In more complex animals, phagocytosis of bacteria and other microorganisms is an essential component of the innate immune response. In addition, phagocytosis is central to the removal of apoptotic bodies during morphogenesis and tissue remodeling. After scission from the plasmalemma, the phagosomal membrane undergoes a major overhaul through a complex series of fission and fusion reactions with endomembrane compartments. This process, collectively called "maturation", confers on the phagosome the ability to kill and degrade the ingested material.

Phagocytosis and phagosomal maturation are elaborate processes which involve signal transduction, cytoskeletal rearrangements, membrane traffic and ion transport, a virtual compendium of cellular physiology which cannot be easily covered by a single review. The limited goal of the present chapter is to summarize our current knowledge of the role of phosphoinositides in phagosome formation and maturation. To this end, a brief primer on phagocytosis and maturation is offered by way of introduction. More detailed accounts on other aspects of phagocytosis can be found in existing reviews (Aderem and Underhill 1999; Deretic and Fratti 1999; Duclos and Desjardins 2000; Franc et al. 1999; Kwiatkowska and Sobota 1999; May 2001; Mayorga et al. 1991; Ofek et al. 1995; Rupper and Cardelli 2001; Tjelle et al. 2000; Vieira et al. 2002).

2
Phagocytosis

Phagocytosis is a receptor-initiated process. Not surprisingly, different receptors trigger particle ingestion in different cell types. However, even in individual cell types, multiple phagocytic receptors can co-exist. Thus mammalian macrophages express at least half a dozen distinct phagocytic receptor types. Phagocytosis can be an opsonin-dependent or -independent process. In the case of non-opsonic engulfment, phagocytes directly recognize endogenous conserved moieties on the surface of the target particle, such as lipopolysaccharide and glycans present in gram-negative bacterial and fungal cell walls, respectively. Scavenger, mannose and β-glycan receptors can mediate such non-opsonic uptake (Kwiatkowska and Sobota 1999). Opsonins are extraneous ligands which coat the target particle, making it more appetizing for ingestion by phagocytes. They are exemplified by complement and IgG. Complement fragments recognize another opsonin, IgM, or promiscuously bind to the surface of pathogens. IgG links the innate and adaptive immune responses by binding to cognate antigens on the surface of invading microorganisms (Kwiatkowska and Sobota 1999). There are a number of complement receptors, CR1 through CR4, of which CR3 and CR4 belong to the integrin family. IgG-opsonized particles are recognized by the Fcγ receptors (FcγR), which bind to the constant domain of IgG. There are three classes of FcγR: FcγRI, FcγRII and FcγRIII, each with subclasses (Gessner et al. 1998).

Not all forms of phagocytosis are identical. The mode of particle internalization and the associated reactions triggered by phagocytosis depend on the receptor engaged. For example, IgG-opsonized particles are engulfed by extension of pseudopodia, whereas complement-coated particles gently "sink" into the phagocytic cell (Allen and Aderem 1996; Kaplan 1977). In addition, engagement of FcγR, but not CR3, induces an inflammatory response which includes cytokine release and generation of reactive oxygen intermediates (Aderem et al. 1985; Wright and Silverstein 1983). Despite these differences, phagocytic receptors rarely operate individually and are more likely to cooperate, as most phagocytic particles expose multiple types of ligands. For simplicity, the remainder of this chapter focuses on FcγRs, which have been most widely studied and are better understood.

Signal transduction leading to phagocytosis is thought to be initiated by phosphorylation of an immunoreceptor tyrosine-based activation motif (ITAM) present on the Fcγ receptor chain or on its ancillary subunits. Src-family kinases, notably Lyn, are responsible for this phosphorylation (Gessner et al. 1998; Ghazizadeh et al. 1994). The resulting phosphotyrosine moieties then function as sites for anchorage of Syk, a tyrosine kinase with dual SH2 domains. Syk, in turn, is believed to phosphorylate and thereby activate a variety of effectors, notably the p85 subunit of phosphatidylinositol 3-kinase (PtdIns 3-kinase), phospholipase Cγ

(PLCγ), Shc and Vav (Jabril-Cuenod et al. 1996; Kwiatkowska and Sobota 1999).

Signals emanating from these and other effectors translate into extensive actin remodeling at the base of the forming phagosome, known as the phagocytic cup. Rho-family GTPases are responsible for the rearrangement of actin, with Rac and Cdc42 believed to be most important in the case of FcγR-mediated phagocytosis (Caron and Hall 1998; Cox et al. 1997; Massol et al. 1998). ARF6 is also involved in phagocytosis, either by activating phosphoinositide kinases and/or by facilitating the focal exocytosis of endomembranes which is thought to contribute to the extension of pseudopods (Bajno et al. 2000; Zhang et al. 1998). Additional small GTPases can also contribute to particle internalization in the case of other receptors. R-Ras and Rap modulate the activation of CR3, which in turn promotes actin polymerization via RhoA (Caron et al. 2000; Self et al. 2001).

The restructuring of actin is coordinated with membrane fusion events which ultimately lead to scission of the phagocytic vacuole from the surface membrane (Bajno et al. 2000). Lipid metabolism is most likely instrumental to the remodeling of the membranes, but the underlying events are poorly understood. Products of phospholipase A_2 (PLA_2), PLC and phospholipase D (PLD), plus ceramide and sphingolipids, have all been implicated directly or indirectly in phagocytosis (Baumruker and Prieschl 2002; Hinkovska-Galcheva et al. 2002; Lennartz 1999; Suchard et al. 1997). These are not considered further here, as emphasis is placed on the phosphoinositides.

3
Phagosome Maturation

Phagocytosis, per se, does not eliminate internalized pathogens. In fact, many pathogens induce their own internalization, effectively invading cells and gaining access to the benign intracellular milieu, hidden from immune sentinels. The luminal contents of newly formed phagosomes are similar to the harmless extracellular space, and their membrane is similar to the plasmalemma. Effective killing and degradation of the internalized pathogen depends on the ensuing phagosome maturation, a process of extensive remodeling of the phagosomal membrane and contents (Vieira et al. 2002).

Although a role for the endoplasmic reticulum was recently invoked (Gagnon et al. 2002), maturation of phagosomes is usually regarded as resulting from a series of fusion reactions with elements of the endocytic pathway. Through these interactions the phagosome acquires vacuolar-type ATPases which acidify its lumen, oxidases which generate free radicals and other toxic metabolites and hydrolytic enzymes which digest proteins and lipids. In parallel with the fusion reactions, fission events serve to recycle MHC-II/antigen complexes for presentation at the cell surface, to remove undesirable components from the phagosomal membrane and to maintain the phagosomal surface and volume near constant throughout maturation (Tjelle et al. 2000; Vieira et al. 2002). Like phagosome formation, maturation is a highly ordered process. Nascent phagosomes have a propensity to fuse with early endosomes yet are refractory to interactions with late endosome/lysosomes. Conversely, late phagosomes interact with lysosomes but have lost the ability to fuse with early endosomes (Desjardins et al. 1994; Desjardins et al. 1997; Jahraus et al. 1998; Mayorga et al. 1991). Thus phagosome maturation is a time-dependent, unidirectional process.

Immediately after closure, and possibly even earlier, the newly formed phagosome initiates interactions with sorting and recycling endosomes (Desjardins et al. 1994). The onset of luminal acidification and the enrichment of the vacuolar membrane with transferrin receptors attest to this occurrence. A unique set of Rab and SNARE proteins, including VAMP2 and/or -3, syntaxin 13 and Rab5, are thought to direct this process (Bajno et al. 2000; Collins et al. 2002; Fratti et al. 2001). Like the early endosome, recently formed phagosomes acquire early endosome autoantigen 1(EEA1) and shortly thereafter also Rab7. There is accumulating evidence that these molecules, with participation of the Rab7 effector RILP (R. Harrison and S. Grinstein, unpublished results), contribute to the subsequent steps in the maturation sequence, although details remain sketchy. Syntaxin 7, which has been detected in late phagosomes, is likely to mediate one of the secondary fusion events, and the presence of lyso*bis*phosphatidic acid (LBPA) suggests that multivesicular bodies may be formed as part of the maturation process (Collins et al. 2002). Eventually, the phagosome merges with lysosomes, yielding a hybrid organelle, the phago-lysosome, which is very acidic and is rich in degradative enzymes and toxic metabolites, creating an environment which is conducive to the elimination of engulfed objects (Desjardins et al. 1994). Although the phago-lysosome is often thought to be the terminal compartment of the maturation pathway, the reality may be more complex. Recently, exocytosis of phago-lysosomes has been reported, suggesting that an additional mechanism may exist for clearance of indigestible material (Raucher et al. 2000).

4
Phosphoinositides

Phosphoinositides (PtdIns) and the products of their metabolism are currently the focus of intense attention by scientists in multiple disciplines, as they appear to play critical roles in signal transduction, regulation of the actin cytoskeleton, membrane fission and fusion and even the control of gene expression. This functional versatility is attributable to the multiplicity of chemical species which can be generated by permutation of the sites phosphorylated on the inositol headgroup. Seven such species are identifiable: phosphatidylinositol-3-phosphate [PtdIns(3)P], phosphatidylinositol-4-phosphate [PtdIns(4)P], phosphatidylinositol-5-phosphate [PtdIns(5)P], phosphatidylinositol-3,4-bisphosphate [PtdIns(3,4)P_2], phosphatidylinositol-3,5-bisphosphate [PtdIns(3,5)P_2], phosphatidylinositol-4,5-bisphosphate [PtdIns(4,5)P_2] and phosphatidylinositol-3,4,5-trisphosphate [PtdIns(3,4,5)P_3] (Cockcroft and De Matteis 2001; Fruman et al. 1998; Payrastre et al. 2001).

Individual phosphoinositides can interact with a unique subset of cellular proteins, thereby conveying structural and/or functional information (Table 1). These interactions are mediated by defined protein domains or motifs which associate with the headgroups of their target phosphoinositides with varying affinity and specificity. The growing list of phosphoinositide-interacting domains includes pleckstrin homology (PH) domains, epsin amino-terminal homology (ENTH) domains, band 4.1-ezrin-radixin-moesin (FERM) domains, Fab1-YOTB-Vac1p-EEA1 (FYVE) domains and Phox domains (PX) (Cullen et al. 2001). There are also a number of heterogeneous polybasic motifs which associate electrostatically with polyphosphoinositides (Laux et al. 2000; McLaughlin et al. 2002; Wang et al. 2002). Although phosphoinositides recruit and anchor proteins to membranes, persistent association often necessitates additional interactions, including contact of the phosphoinositide-binding protein with other proteins in the vicinity of the cognate lipid (Fushman et al. 1998; Varnai et al. 2002). Nevertheless, several domains bind defined phosphoinositides with high affinity and specificity even when

Table 1 Synthesis, consumption and effectors of phosphoinositides and diacylglycerol

Lipid	Enzyme		Substrate	Product	Effectors
$PI(4,5)P_2$	Synthesis	PIPKI	$PI(4)P$		Dynamin-2;
		PIPKII	$PI(5)P$		Actin binding proteins: gelsolin, profilin cofilin,
		PTEN	$PI(3,4,5)P_2$		WASP, talin, vinculin, ERM
	Consumption	PLC		$DAG + IP_3$	
		5'Pases		$PI(4)P$	
		4'Pases		$PI(5)P$	
		PI3Ks		$PI(3,4,5)P_3$	
DAG	Synthesis	PLC	$PI(4,5)P_2$		PKCs
		PLD/PAP	PC/PA		RasGRPs
	Consumption	DAGK		PA	Chimaerins
$PI(3,X)P_{Poly}$	Synthesis	PI3Ks	$PI(4)P$ or $PI(4,5)P_2$		Myosin X
					ARF6 GEFs
	Consumption	PTEN		$PI(4)P$	Rho GEFs
				$PI(4,5)P_2$	AKT, aPKCs
		SHIP		$PI(3,4)P_2$	
$PI(3)P$	Synthesis	VPS34	PI		EEA1, Hrs
	Consumption	3'-Pases		PI	NADPH oxidase
		PIKfyve		$PI(3,5)P_2$	PIKfyve
					Sorting nexins

excised from the originating protein. Indeed, the affinity and selectivity of the interaction are sufficiently high to warrant their use as probes to monitor the presence and distribution of their target phosphoinositide. Constructs encoding GFP chimeras of the FYVE domain of EEA1, or of the PH domains of Grp1 or PLCδ, have been used extensively to detect PtdIns(3)P, PtdIns(3,4,5)P$_3$ and PtdIns(4,5)P$_2$, respectively, in intact cells (Klarlund et al. 1997; Kutateladze et al. 1999; Stauffer et al. 1998).

The remainder of this chapter summarizes our current knowledge of the involvement of PtdIns(4,5)P$_2$, PtdIns(3,4)P$_2$, PtdIns(3,4,5)P$_3$ and PtdIns(3)P in phagocytosis and phagosome maturation.

5
Phosphatidylinositol-4,5-bisphosphate

As in other cells, PtdIns(4,5)P$_2$ in quiescent phagocytes distributes predominantly to the cytosolic leaflet of the plasma membrane and the Golgi network (Godi et al. 1999; Verkleij and Post 2000). PtdIns(4,5)P$_2$ is primarily synthesized by phosphorylation of PtdIns(4)P at the 5′ position by type I phosphoinositide phosphate kinases (PIPKI) (Rameh et al. 1997). Other pathways, such as the phosphorylation of PtdIns(5)P by type II phosphoinositide phosphate kinases (PIPKII) (Rameh et al. 1997) or the dephosphorylation of PtdIns(3,4,5)P$_2$ by the 3′-phosphatase PTEN (Maehama and Dixon 1998), may also contribute to the formation of PtdIns(4,5)P$_2$, but the magnitude of this contribution is ill defined.

There are three PIPKI isoforms in mammalian cells: α, β and γ (the nomenclature for the human and murine α and β isoforms is, unfortunately, reversed). PIPKI isoforms have been detected in the cytosol or bound to the plasma or Golgi membranes, as well as associated with filamentous actin and other uncharacterized internal structures (Brown et al. 2001; Divecha et al. 2000; Hinchliffe et al. 2002; Jones et al. 2000; Kunz et al. 2000). Importantly, both the subcellular distribution and the catalytic activity of PIPKI are regulated by GTPases of the Rho (e.g., Rac) and ARF families (e.g., ARF6), as well as by phosphatidic acid, a product of PLD (Honda et al. 1999; Tolias et al. 2000). Interestingly, optimal PLD activity itself requires the presence of PtdIns(4,5)P$_2$, suggesting the existence of an amplifying, positive feedback loop (Divecha et al. 2000; Honda et al. 1999; Jenkins et al. 1994).

The concentration of PtdIns(4,5)P$_2$ can be decreased by the action of kinases, phosphatases or lipases. The phosphoinositide 5-phosphatases which consume PtdIns(4,5)P_2, exemplified by the synaptojanins, contain catalytic domains which have been conserved from yeast to mammals, and often also Sac domains. Comparatively little is known regarding the regulation of phosphoinositide 5-phosphatases (Hughes et al. 2000; Sakisaka et al. 1997). In contrast, a great deal has been learned about the conversion of PtdIns(4,5)P_2 to PtdIns(3,4,5)P_3 by PtdIns 3-kinases and about its cleavage by PLC isoforms, which yields diacylglycerol (DAG) and inositol 3,4,5-*tris*phosphate (IP$_3$). These enzymes are discussed in more detail below in the context of phagosome formation and maturation.

In addition to providing a substrate for the formation of important messengers such as DAG, IP$_3$ and PtdIns(3,4,5)P_3, PtdIns(4,5)P_2 itself dictates the distribution and modulates the activity of a vast number of proteins with PH, FERM, ENTH or VHS domains, or polybasic motifs which bind to its headgroup (Cullen et al. 2001). Such proteins in all likelihood play a central role in phagosome formation. Indeed, acute changes in the local concentration of PtdIns(4,5)P_2 have been reported to occur during particle engulfment. With the use of the PH domain of PLCδ fused to GFP, PtdIns(4,5)P_2 was found to undergo a biphasic change at sites of phagocytosis (Botelho et al. 2000). During the initial stages, PtdIns(4,5)P_2 was observed to accumulate in extending pseudopods relative to other areas of the plasma membrane. Subsequently, the concentration of PtdIns(4,5)P_2 decreased rapidly, disappearing from the center of the phagocytic cup even before phagosomal closure. By the time it had sealed and severed from the plasmalemma, the phagosome had no detectable PtdIns(4,5)P_2 left on its membrane (Figs. 1A and 2).

The early phase of PtdIns(4,5)P_2 accumulation in the pseudopods is suggestive of localized stimulation of synthesis of the phosphoinositide. Indeed, recent work by Coppolino et al. (2002) provided evidence that PIPKIα is preferentially recruited to the phagosomal cup at early stages of the ingestion process. Moreover, these authors showed that although catalytically inactive mutants of PIPKIα did not alter the basal distribution of PtdIns(4,5)P_2 and actin, they interfered with the focal accumulation of PtdIns(4,5)P_2 and the associated deposition of actin normally seen during phagocytosis (Coppolino et al. 2002). How PIPKI is recruited and regulated during phagocytosis remains to be elucidated, but it is likely that ARF6, Rac and/or PLD act in concert to stimulate PtdIns(4,5)P_2 generation. It is noteworthy, in this regard, that both types of GTPases and PLD are known to be activated during and essential for

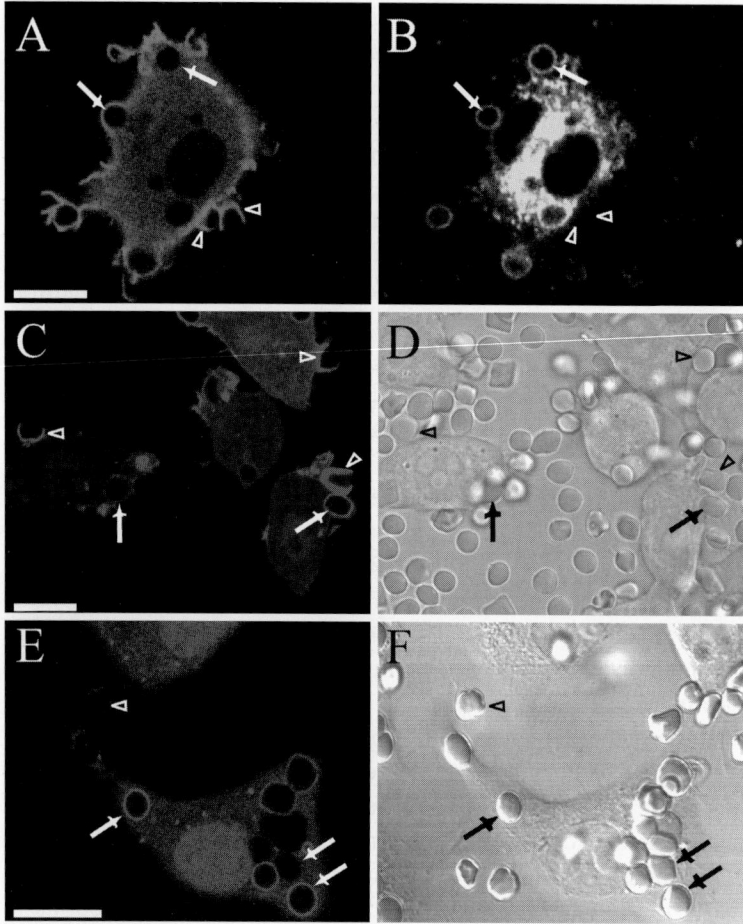

Fig. 1A–F Visualization of phosphoinositides during phagocytosis and phagosome maturation. Macrophage-like RAW264.7 cells were transfected with fluorescent phosphoinositide-binding probes and allowed to internalize IgG-opsonized sheep red blood cells. **A, B** Cells co-expressing a fusion of CFP with the PH domain of PLCδ (PLCδ-PH-CFP), a probe for PtdIns(4,5)P$_2$ (**A**), and a fusion of YFP with the C1 domain of PKCδ (C1δ-YFP), a probe for DAG (**B**). In these and subsequent panels phagocytic cups are noted by *open arrowheads* and sealed phagosomes by *closed arrows*. PLCδ-PH-CFP [i.e., PtdIns(4,5)P$_2$] undergoes a transient accumulation in the early phagocytic cups but is completely absent from formed phagosomes. C1δ-YFP (i.e., DAG) is barely detectable in early phagocytic cups but can be clearly seen in recently formed phagosomes. **C, D** Macrophages expressing the PH domain of Akt fused to GFP (Akt-PH-GFP), a probe for 3″-polyphosphoinositides. A fluorescence image is shown in **C**, and the corresponding differential interference contrast image is shown

effective phagocytosis (Cox et al. 1997; Kusner et al. 1999; Massol et al. 1998; Zhang et al. 1998).

The precise role which PtdIns(4,5)P_2 plays in the early stages of phagocytosis is unclear, but modulation of the polymerization of actin is a very likely candidate. A wealth of literature links PtdIns(4,5)P_2 to the regulation of the actin cytoskeleton (see the chapter by Hilpelä et al., this volume). Generally, PtdIns(4,5)P_2 increases the number of free barbed ends available in actin filaments, spurring polymerization. This is accomplished in part by removal of capping proteins, such as CapZ, CapG and gelsolin, from barbed ends (Heiss and Cooper 1991; Witke et al. 2001; Yin et al. 1981). Of note, gelsolin has been reported to accumulate on forming phagosomes, and phagocytosis is defective in neutrophils from gelsolin$^{-/-}$ mice (Serrander et al. 2000; Yin et al. 1981). More recently, CapG-deficient macrophages were similarly found to have impaired phagocytosis (De Corte et al. 1997; Witke et al. 2001). Alternatively, new barbed ends can be generated either by severing existing filaments, a reaction catalyzed by members of the ADF/cofilin family, or by de novo assembly of actin filaments, mediated by the ARP2/3 complex (see below; Takenawa and Itoh 2001). Consistent with a role in phagocytosis, microinjection of anti-cofilin antibodies attenuated phagocytosis in neutrophils (Nagaishi et al. 1999). However, in apparent contradiction with these findings, depletion of cofilin by antisense technology augmented phagocytosis in macrophages (Adachi et al. 2002). Finally, PtdIns(4,5)P_2 can promote de novo actin polymerization through N-WASP (Wiskott-Aldrich syndrome protein). Acting in conjunction with activated Cdc42, PtdIns(4,5)P_2 promotes the disinhibition of N-WASP, allowing it to bind and activate the ARP2/3 complex, which in turn serves as a nucleus for the emergence of new actin filaments (Higgs and Pollard 2000; Prehoda et al. 2000; Rohatgi et al. 2000). Both WASP and

◄─────────────────────────────────────

in D. There is a substantial recruitment of Akt-PH-GFP (i.e., 3′-polyphosphoinositides) to phagocytic cups and recently formed phagosomes but not to more mature phagosomes. E, F Macrophages expressing GFP fused to two tandem FYVE domains of EEA1 (2FYVE-GFP; E), a probe for PtdIns(3)P. A fluorescence image is shown in E, and the corresponding differential interference contrast image is shown in F. Recruitment of 2FYVE-GFP [i.e., PtdIns(3)P] occurs only after phagosome closure and is transient, lasting 5–10 min. Scale bars=10 µm

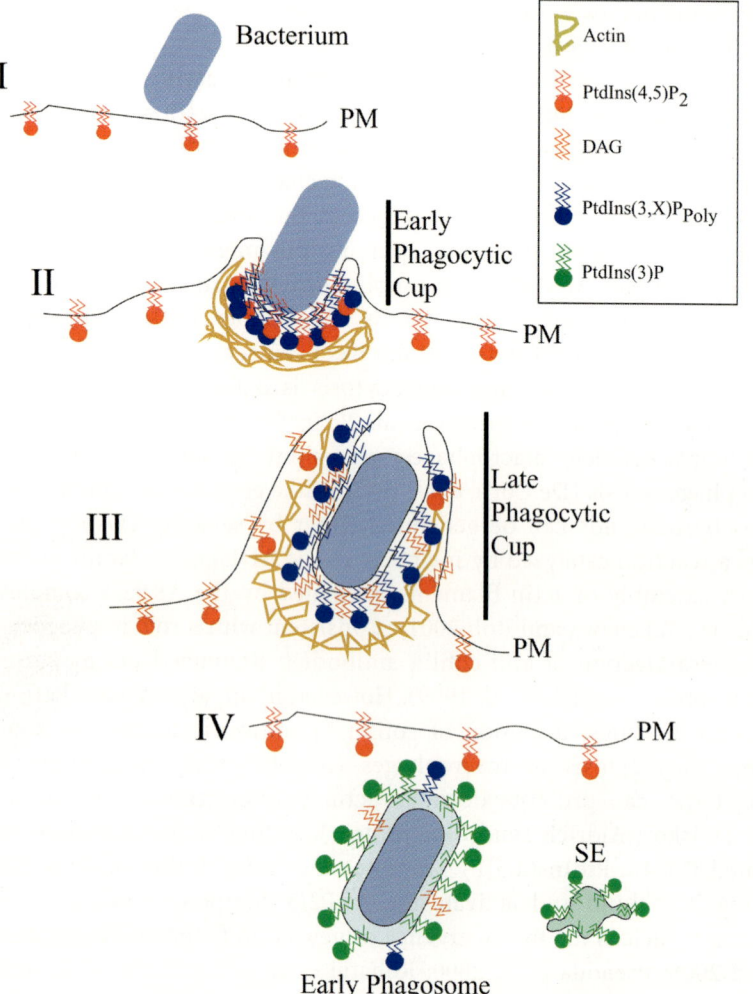

Fig. 2 Diagram illustrating phosphoinositide metabolism during phagocytosis and phagosome maturation. *I*: The plasma membrane (*PM*) of a resting macrophage contains PtdIns(4,5)P$_2$ but little DAG, 3'-polyphosphoinositides [PI(3,X)$_{poly}$] or PtdIns(3)P. *II*: On engagement of phagocytic receptors, PtdIns(4,5)P$_2$ and 3'-polyphosphoinositide synthesis is stimulated, leading to localized actin polymerization. *III*: As phagocytosis proceeds, PtdIns(4,5)P$_2$ disappears from phagocytic cups as they near closure, presumably because of activation of PLCγ and PtdIns 3-kinases. Accordingly, DAG can be detected at this stage and in recently formed phagosomes. *IV*: After phagosome closure, PtdIns(4,5)P$_2$ and actin are no longer detectable on the phagosomal membrane. DAG and 3'-polyphosphoinositides disappear within 1–2 min of phagosome sealing. At this time, PtdIns(3)P is found to accumulate on the early phagosome, where it persists for 5–10 min. PtdIns(3)P is implicated in fusion of phagosomes with endocytic compartments, including the sorting endosomes (*SE*)

ARP2/3 are enriched in phagocytic structures and are required for their formation (Lorenzi et al. 2000; May et al. 2000).

Conceivably, PtdIns(4,5)P_2 can also impact on particle ingestion by modulating the anchorage of actin filaments to the plasma membrane. Indeed, PtdIns(4,5)P_2 regulates the localization and activity of a number of proteins, including talin, vinculin and α-actinin, which are thought to link F-actin to the cytosolic aspect of the membrane. In this regard, it is noteworthy that all three of these linker proteins were found to concentrate on forming phagosomes. In summary, although it has yet to be tested directly, PtdIns(4,5)P_2 likely operates in phagocytosis by regulating actin remodeling.

At later stages of phagosome formation, PtdIns(4,5)P_2 can no longer be detected by the PH domain of PLCδ (Botelho et al. 2000) (Figs. 1A and 2). At least two mechanisms contribute to the apparent consumption of PtdIns(4,5)P_2: conversion to PtdIns(3,4,5)P_3 by PtdIns 3-kinase (see below) and hydrolysis by PLCγ. The products of hydrolysis of PtdIns(4,5)P_2 by PLCγ are discussed in detail below and an entire section is devoted to the formation and role of PtdIns(3,4,5)P_3 in phagocytosis. Before discussing the relevance of the products of PtdIns(4,5)P_2 metabolism, however, we should consider the consequences of the disappearance of the substrate itself. To the extent that PtdIns(4,5)P_2 is required for the anchorage and expansion of the actin meshwork underlying the phagosomal cup, its conversion to other species will likely result in diminution of actin accumulation. Indeed, it has been shown with laser tweezers and atomic force microscopy that elimination of PtdIns(4,5)P_2 markedly reduces membrane rigidity, indicative of weakening of the actin skeleton which supports the membrane (Raucher et al. 2000; Terebiznik et al. 2002). Hence, a very alluring possibility is that loss of PtdIns(4,5)P_2 may itself signal disassembly of the actin which underlies the cup, facilitating phagosome closure and detachment from the plasmalemma. In this event, synthesis and elimination of PtdIns(4,5)P_2 would have to be carefully coordinated in time and space to allow sequential actin assembly and disassembly.

Numerous studies have documented increased liberation of DAG during phagocytosis (Della Bianca et al. 1993; Della Bianca et al. 1991; Della Bianca et al. 1990; Fallman et al. 1992; Fallman et al. 1989; O'Shea et al. 1985). The precise kinetics and site of generation of DAG during phagocytosis was visualized with a fluorescent chimera of the C1 domain of PKCδ, which selectively binds to this product of PtdIns(4,5)P_2 hydrolysis

(Botelho et al. 2000; Oancea et al. 1998). Remarkably, DAG was found to form at the time and in the precise region where PtdIns(4,5)P$_2$ was disappearing, implicating PLC in the conversion (Figs. 1B and 2). Accordingly, liberation of IP$_3$ and increased cytosolic calcium have also been reported to occur at the early stages of phagocytosis (Della Bianca et al. 1993; Fallman et al. 1989) and PLCγ was found to be recruited to the phagosome (Botelho 2000). Although DAG is undoubtedly generated by PLCγ during phagocytosis, other sources also exist. A fraction of the DAG generated by activation of phagocytic receptors is derived from phosphatidic acid, which is generated by PLD and is converted to DAG by a phosphatase (Della Bianca et al. 1991; Fallman et al. 1992).

The functional significance of DAG to particle ingestion is still being investigated. Exogenous addition of DAG or of phorbol esters which mimic its action increases phagocytic efficiency (Karimi et al. 1999; Larsen et al. 2000). This augmentation has been attributed to activation of classical or novel PKC isoforms. Indeed, PKCα, PKCδ and PKCε are recruited to phagosomes formed by FcγR clustering in macrophages (Brumell et al. 1999, Larsen et al. 2000; R. Botelho and S. Grinstein, unpublished observations). However, although some investigators report that FcγR-mediated phagocytosis requires PKC (Andersson et al. 1988; Karimi et al. 1999; Newman et al. 1991; Zheleznyak and Brown 1992), others disagree (Larsen et al. 2000; Newman et al. 1991).

The potential effects of DAG in phagocytosis are not restricted to the activation of PKC. Other potential targets are the chimaerins, a family of Rac GTPase-activating proteins, and members of the RasGRP family of Ras guanine nucleotide exchange factors (Ahmed et al. 1990; Caloca et al. 1997; Caloca et al. 1999; Ebinu et al. 1998; Tognon et al. 1998) (Table 1). Both chimaerins and RasGRP possess C1 domains capable of binding to DAG. By terminating the action of Rac, chimaerins could conceivably be engaged in actin remodeling during particle engulfment. Complexation of DAG by RasGRP isoforms would induce Ras activation at the phagosome. We have preliminary evidence that RasGRP3 translocates to phagocytic cups (R. Botelho, unpublished observations), bolstering this possibility. Hence, the DAG-dependent augmentation of phagocytosis may proceed, at least in part, through pathways not involving PKC.

6 Phosphatidylinositol-3,4-bisphosphate and Phosphatidylinositol-3,4,5-trisphosphate

PtdIns(3,4)P_2 and PtdIns(3,4,5)P_3 are found in negligible amounts in resting cells but are quickly generated on stimulation with various agonists (Vanhaesebroeck and Waterfield 1999). Synthesis appears to be mostly restricted to the plasma membrane, although there is evidence that 3′-polyphosphoinositides may regulate intracellular membrane traffic (Christoforidis et al. 1999; Kurosu and Katada 2001).

Synthesis of PtdIns(3,4)P_2 and PtdIns(3,4,5)P_3 is predominantly realized by phosphorylation of PtdIns(4)P and PtdIns(4,5)P_2, respectively, by PtdIns 3-kinases, of which there are three classes (Vanhaesebroeck and Waterfield 1999). Only class I and class II kinases can generate 3′-polyphosphoinositides [see below for class III and its product PtdIns(3)P]. Class I is further subdivided into class IA and class IB, characterized by a regulatory p85 subunit (an alternative p55 subunit also exists) which responds to tyrosine kinases and a regulatory p101 subunit which responds to trimeric G-proteins, respectively. Both p85 and p101 regulatory subunits modulate the activity and stability of the catalytic p110 subunits (Vanhaesebroeck and Waterfield 1999). Class II kinases are composed of a single polypeptide containing a C2 domain which allows interaction with phospholipids (Vanhaesebroeck and Waterfield 1999). PtdIns(3,4)P_2 can also be generated by hydrolysis of the 5′ phosphate of PtdIns(3,4,5)P_3, a reaction catalyzed by the phosphatase SHIP. Disappearance of PtdIns(3,4,5)P_3 can also result from the activity of PTEN, which dephosphorylates the 3′ position of polyphosphoinositides, including PtdIns(3,4)P_2.

Class I PtdIns 3-kinases are recruited and activated at sites of phagocytosis (Marshall et al. 2001). Moreover, several lines of evidence indicate that class I PtdIns 3-kinases are required for efficient FcγR-mediated phagocytosis. First, antagonists of PtdIns 3-kinases impede phagocytosis in macrophages, affecting larger particles more than smaller ones (Araki et al. 1996; Cox et al. 1999; Ninomiya et al. 1994). Second, cells from p85$\alpha^{-/-}$p85$\beta^{-/-}$ knockout mice internalize large particles much less effectively than cells from wild-type littermates (Vieira et al. 2001). These studies were performed in FcγR-transfected embryonic fibroblasts, because the double-knockout animals were not viable, precluding the isolation of primary phagocytes. Third, cross-linking of a transmembrane chimera bearing the p85 subunit of the class I kinase on its cytosolic domain sufficed to induce phagocytosis (Lowry et al. 1998). Finally, overexpression of SHIP attenuates phagocytosis, suggesting that PtdIns(3,4,5)P_3, rather than PtdIns(3,4)P_2, is key for phagocytosis (Cox et al. 2001).

PtdIns 3-kinases would seem at first glance to be unsuitable to regulate phagocytosis, a spatially restricted process, because their lipid products are predicted to diffuse rapidly along the fluid mosaic of the membrane. Such lateral diffusion would delocalize the signal, which appears incompatible with the highly restricted nature of the phagocytic event. However, direct observations with confocal microscopy revealed that the PtdIns(3,4,5)P_3 and/or PtdIns(3,4)P_2 formed during ingestion were exquisitely confined to the nascent phagosome (Marshall et al. 2001) (Figs. 1C and 2). The factor(s) which confine the 3′-polyphosphoinositides to the cup remain undefined, but immobilization by rafts or trapping by interaction with poorly mobile protein ligands are logical possibilities. It is also possible that PtdIns(3,4,5)P_3 is continuously generated at sites of phagocytosis, diffuses laterally and is degraded at the edges of the cup by active phosphatases, yielding a stationary gradient of the lipid, despite the mobility of individual molecules.

As intimated above, inhibition of particle engulfment by PtdIns 3-kinase antagonists is proportional to the diameter of the target particle, suggesting a size-dependent threshold for the PtdIns 3-kinase requirement in phagocytosis (Cox et al. 1999; Vieira et al. 2001). PtdIns 3-kinases seem to be involved in pseudopod elongation, because cells treated with wortmannin show blunted pseudopod extension (Cox et al. 1999). The greater demand for membrane elongation explains the preferential inhibition of large-particle uptake. Because actin still accumulates under the blunted phagocytic cups formed in wortmannin-treated cells, it is likely that PtdIns 3-kinase functions instead in the local fusion of membranes which facilitates pseudopod extension. However, the possibility that PtdIns 3-kinase is necessary for some aspect of long-range actin remodeling cannot be dismissed. In fact, myosin X, which possesses three PH domains that interact with PtdIns(3,4,5)P_3, was shown recently to be essential for particle internalization (Cox et al. 2002). The PH domains are critical for myosin X recruitment to nascent phagosomes and for completion of particle uptake. These findings are in accord with earlier observations indicating that contractile activity is essential to complete phagocytosis (Swanson et al. 1999).

There are also several guanine nucleotide exchange factors which bear PH domains and are regulated by PtdIns(3,4,5)P$_3$ (Han et al. 1998; Stephens et al. 2002). Grp1, ARNO and cytohesin-1 are activators of ARF proteins, including ARF6, which is required for particle ingestion (Klarlund et al. 1997; Venkateswarlu et al. 1998; Zhang et al. 1998). As discussed above, ARF6 is implicated in PtdIns(4,5)P$_2$ synthesis, and therefore in actin assembly, but also in the regulation of membrane traffic. Either of these functions could account for the PtdIns(3,4,5)P$_3$ requirement of phagocytosis.

In summary, PtdIns(3,4,5)P$_3$ [and possibly also PtdIns(3,4)P$_2$] are formed transiently during the early stages of phagocytosis. They accumulate at the cup and persist for a very short time on the membrane of the phagosome after it detaches from the surface. The 3′-polyphosphoinositides are essential for optimal internalization of large (3 ≧ μm) particles but are dispensable for ingestion of smaller ones. This dependence likely involves various downstream effectors, including myosin X, exchange factors which activate ARF6 and in all likelihood other effectors which were heretofore unrecognized.

7
Phosphatidylinositol-3-phosphate

In mammalian cells, PtdIns(3)P is most abundant in endosomes (Gillooly et al. 2000). Endosomal PtdIns(3)P is generated by the class III PtdIns-3 kinase hVPS34, which preferentially phosphorylates the 3′ position of phosphatidylinositol (Volinia et al. 1995). Like the class I kinases, hVPS34 is sensitive to low concentrations of wortmannin, and exposure to this fungal metabolite or microinjection of inhibitory anti-hVPS34 antibodies results in rapid depletion of endosomal PtdIns(3)P (Stephens et al. 1994). Recruitment of hVPS34 to endosomes is mediated by the adaptor protein p150, a myristoylated serine/threonine kinase. Association with p150, which is thought to be a Rab5 effector protein (Murray et al. 2002), is required for catalytic activation of hVPS34.

Less is known about the catabolism of PtdIns(3)P. Theoretically, loss of PtdIns(3)P can occur through at least three different mechanisms: dephosphorylation of the 3′ phosphate, yielding PtdIns, conversion of PtdIns(3)P into a *bis*phosphate species through the action of PtdIns kinases or degradation of the phosphoinositide by phospholipase activity. Recently, candidate molecules for each of these functions have been described; however, their relative contribution to the removal of PtdIns(3)P in vivo is not clear. Two proteins which display PtdIns(3)P-specific phosphatase activity, namely FYVE-DSP1 and FYVE-DSP2, were recently identified (Zhao et al. 2001). Furthermore, PIKfyve, the mammalian homologue of the yeast protein Fab1, is a phosphoinositide 5′-kinase which can convert PtdIns(3)P to PtdIns(3,5)P_2 (Shisheva 2001). Finally, several phospholipases have been shown to hydrolyze PtdIns(3)P, including some PtdIns-PLD isozymes which were found to prefer PtdIns(3)P as a substrate in vitro (Ching et al. 1999). More work will be required to determine which of these enzymes contribute to the catabolism of PtdIns(3)P in situ.

Our understanding of the functional role of PtdIns(3)P has been greatly advanced by the identification of two distinct protein domains which interact selectively with PtdIns(3)P: the FYVE domain (Burd and Emr 1998) and the Phox or PX domain (Cheever et al. 2001; Ellson et al. 2001; Song et al. 2001; Xu et al. 2001; Yu and Lemmon 2001). Expression of fluorescent fusion proteins incorporating these PtdIns(3)P-binding modules has allowed visualization of the distribution of the phosphoinositide in living cells. As mentioned above, PtdIns(3)P is primarily localized to sorting endosomes (Fig. 2) and multi-vesicular bodies but is also detectable in the nucleolus (Gillooly et al. 2000).

In contrast to the other phosphoinositides discussed above, PtdIns(3)P is thought to function not so much as a signaling molecule, but rather as a participant in cellular housekeeping functions. PtdIns(3)P and PtdIns(3)P-interacting proteins play critical regulatory roles in the control of membrane traffic in diverse cellular processes, including endosome fusion (Simonsen et al. 1998), autophagy (Petiot et al. 2000; Wurmser and Emr 2002), growth factor receptor processing (Itoh et al. 2002; Oldham et al. 2002) and nuclear membrane dynamics (Roggo et al. 2002). The ability of PtdIns(3)P to participate in such an array of important functions is attributable to the large number of proteins which express either FYVE or PX domains, in combination with various other catalytic or protein interaction domains. These include tethering proteins like EEA1, GTPase effectors like Rabenosyn-5, kinases like PIKfyve and adaptors like Hrs (Stenmark et al. 2002).

PtdIns(3)P was recently appreciated to have a critical role in phagosome maturation. Experiments using chimeras of fluorescent proteins with FYVE or PX domains documented the transient appearance of PtdIns(3)P on the phagosomal membrane, commencing shortly after phagosome closure (Ellson et al. 2001; Vieira et al. 2001) (Figs. 1E and 2). This accumulation was obliterated by pretreatment of the cells with wortmannin. Because this inhibitor eliminates the activity of both the class I and class III PtdIns-3 kinases, these experiments are insufficient to identify the enzyme responsible for the appearance of PtdIns(3)P in phagosomes. To better define the source of PtdIns(3)P, Vieira et al. (2001) used embryonic fibroblasts lacking both the α and β isoforms of the p85 subunit of the class I kinases (see above). In these cells, which retained their ability to engulf small particles, PtdIns(3)P accumulation on phagosomes was seemingly normal, suggesting that class III and not class I kinases were involved. This conclusion was buttressed by microinjection experiments, in which hVPS34 neutralizing antibodies were found to preclude PtdIns(3)P accumulation in phagosomes but not their formation. Accordingly, hVPS34 was detected in phagosomes by immunostaining (Vieira et al. 2001) and also by immunoblotting (Fratti et al. 2001; R. Botelho, unpublished observations). Jointly, these observations indicate that PtdIns(3)P is not apparent in the nascent phagosome and is not required for particle engulfment but that it appears shortly after sealing, through the action of class III PtdIns 3-kinases.

Although inhibition of hVPS34 had no effect on phagosome formation, phagosomal maturation was profoundly altered. In cells treated with wortmannin, or injected with antibodies which neutralize the class III kinase, maturation was arrested at an early stage. The acquisition of EEA1 and LAMP-1 was depressed, and virtually no LBPA was observed (Fratti et al. 2001; Vieira et al. 2001). The precise molecular target responsible for the disruption of maturation has not been fully identified, but EEA1 is a likely candidate. Optimal association of this tethering molecule with phagosomes and endosomes requires interaction of its FYVE domain with PtdIns(3)P on the membrane (Lawe et al. 2002; Simonsen et al. 1998). Failure to recruit EEA1 at the appropriate concentration and/or in the normal configuration may prevent critical fusion events which are indispensable for progression of maturation. Indeed, neutralization of EEA1 function with antibodies resulted in an impairment in maturation which resembled the phenotype obtained in wortmannin-treated cells (Fratti et al. 2001).

Although available evidence suggests that EEA1 is important in phagosomal maturation, the involvement of additional ligands of PtdIns(3)P in the process has not been ruled out and is indeed quite likely. One attractive candidate is Rabenosyn-5, a Rab5 effector protein.

Rabenosyn-5 is recruited by a PtdIns(3)P- and hVPS34-dependent process to early endosomes (Nielsen et al. 2000), where it is thought to direct membrane traffic (de Renzis et al. 2002; Nielsen et al. 2000). Although no evidence is available yet that Rabenosyn-5 activity is required for phagosome maturation, the participation of this protein in endosome progression suggests that it may play an analogous role in phagosome maturation.

At least one other FYVE domain-containing protein associates with early phagosomes, where it interacts with PtdIns(3)P. Unpublished observations from our laboratory, in collaboration with that of H. Stenmark, indicate that Hrs interacts transiently with phagosomes, with a time course which resembles the lifetime of PtdIns(3)P on the vacuole. In addition to the phosphoinositide-binding FYVE domain, Hrs contains other domains and motifs which interact with a bewildering array of partners including ubiquitinated proteins, the R–SNARE SNAP-25, signal-transducing adapter molecule (STAM), the receptor-signaling intermediate Smad2, the clathrin-mediated endocytosis factor Eps15, clathrin itself, the actin-binding protein Schwannomin and sorting nexins (Raiborg et al. 2001a,b). One or more of these may very well influence the fate of phagosomes, and our preliminary experiments show that maturation is arrested at an early phase in cells depleted of Hrs.

Proteins with PX domains are also likely to attach to phagosomes, possibly contributing to maturation. Indeed, the components of the NADPH oxidase which gave their name to PX domains are found in maturing phagosomes, where they form part of the complex which generates microbicidal oxygen radicals (Jankowski and Grinstein 2002). In addition, members of the sorting nexin family are recruited to phagosomes (R. Botelho, unpublished observations). Sorting nexins are a group of at least 27 related proteins, several of which contain PX domains capable of interacting with PtdIns(3)P (Xu et al. 2001). For the most part, their functional role remains nebulous at this time, but there are indications that they contribute to delivery of receptors to the lysosome (Haft et al. 1998; Xu et al. 2001). The role of one member of the family, Sorting Nexin 3 (SNX3), has been examined in some detail. SNX3 resides primarily in the early endosome, where it contributes to epidermal growth factor degradation and transferrin recycling (Xu et al. 2001). If nexins are confirmed to assist in the traffic of membranes, their presence in phagosomes would be suggestive of a role in maturation.

It is becoming increasingly evident that the recruitment of proteins by PtdIns(3)P and the reactions that these proteins initiate are critical for the transition between early and late phagosomes, and possibly for other steps as well. The biological relevance of these events is highlighted by the ability of some microorganisms to disrupt phagosomal maturation. By preventing fusion of phagosomes with lysosomes, these organisms evade the microbicidal response of phagocytes, managing to survive as intracellular parasites within the protected confines of incompletely mature vacuoles. In the case of mycobacteria, Fratti et al. (2001) attributed the maturation arrest to an inability of the phagosomes to recruit EEA1, which is seemingly necessary for the transition to phago-lysosomes. This may be due to the inability of mycobacteria-containing phagosomes to accumulate PtdIns(3)P or, as Fratti and colleagues suggested (Fratti et al. 2001), to the ability of bacterial surface molecules like lipoarabinomannan to interfere with the association between EEA1 and phagosomal PtdIns(3)P.

In summary, although many gaps in our understanding remain to be filled, it is clear that phosphoinositides play multiple roles in the formation, fission and maturation of phagosomes. Moreover, there are initial indications that the generation of inositides and/or their interaction with cellular ligands may be targeted by invading microorganisms and intracellular parasites to subvert the function of host cells. As such, the study of phosphoinositide metabolism acquires an important clinical dimension.

Acknowledgements. Research in the authors' laboratory is supported by the Canadian Institutes of Health Research (CIHR), the Canadian Arthritis Society and the National Sanatorium Association. R.J.B. and C.C.S. are supported by CIHR Graduate Studentships. S.G. is a CIHR Distinguished Scientist and the current holder of the Pitblado Chair in Cell Biology at The Hospital for Sick Children.

References

Adachi R, Takeuchi K, and Suzuki K (2002) Antisense oligonucleotide to cofilin enhances respiratory burst and phagocytosis in opsonized zymosan-stimulated mouse macrophage J774.1 cells. J Biol Chem 23:23

Aderem A, and Underhill DM (1999) Mechanisms of phagocytosis in macrophages. Annu Rev Immunol 17:593–623

Aderem AA, Wright SD, Silverstein SC, and Cohn ZA. (1985) Ligated complement receptors do not activate the arachidonic acid cascade in resident peritoneal macrophages. J Exp Med 161:617–622

Ahmed S, Kozma R, Monfries C, Hall C, Lim HH, Smith P, and Lim L (1990) Human brain n-chimaerin cDNA encodes a novel phorbol ester receptor. Biochem J 272:767–773

Allen LA, and Aderem A (1996) Molecular definition of distinct cytoskeletal structures involved in complement- and Fc receptor-mediated phagocytosis in macrophages. J Exp Med 184:627–637

Andersson T, Fallman M, Lew DP, and Stendahl O (1988) Does protein kinase C control receptor-mediated phagocytosis in human neutrophils? FEBS Lett 239:371–375

Araki N, Johnson MT, and Swanson JA (1996) A role for phosphoinositide 3-kinase in the completion of macropinocytosis and phagocytosis by macrophages. J Cell Biol 135:1249–1260

Bajno L, Peng XR, Schreiber AD, Moore HP, Trimble WS, and Grinstein S (2000) Focal exocytosis of VAMP3-containing vesicles at sites of phagosome formation. J Cell Biol 149:697–706

Baumruker T, and Prieschl EE (2002) Sphingolipids and the regulation of the immune response. Semin Immunol 14:57–63

Botelho RJ, Teruel M, Dierckman R, Anderson R, Wells A, York JD, Meyer T, and Grinstein S (2000) Localized biphasic changes in phosphatidylinositol-4,5-bisphosphate at sites of phagocytosis. J Cell Biol 151:1353–1368

Brown FD, Rozelle AL, Yin HL, Balla T, and Donaldson JG (2001) Phosphatidylinositol 4,5-bisphosphate and Arf6-regulated membrane traffic. J Cell Biol 154:1007–1017

Brumell JH, Howard JC, Craig K, Grinstein S, Schreiber AD, and Tyers M (1999) Expression of the protein kinase C substrate pleckstrin in macrophages: association with phagosomal membranes. J Immunol 163:3388–3395

Burd CG, and Emr SD (1998) Phosphatidylinositol-(3)-phosphate signaling mediated by specific binding to RING FYVE domains. Mol Cell 2:157–162

Caloca MJ, Fernandez N, Lewin NE, Ching D, Modali R, Blumberg PM, and Kazanietz MG (1997) β2-Chimaerin is a high affinity receptor for the phorbol ester tumor promoters. J Biol Chem 272:26488–26496

Caloca MJ, Garcia-Bermejo ML, Blumberg PM, Lewin NE, Kremmer E, Mischak H, Wang S, Nacro K, Bienfait B, Marquez VE, and Kazanietz MG (1999) β2-Chimaerin is a novel target for diacylglycerol: binding properties and changes in subcellular localization mediated by ligand binding to its C1 domain. Proc Natl Acad Sci USA 96:11854–11859

Caron E, and Hall A (1998) Identification of two distinct mechanisms of phagocytosis controlled by different Rho GTPases. Science 282:1717–1721

Caron E, Self AJ, and Hall A (2000) The GTPase Rap1 controls functional activation of macrophage integrin $\alpha_M\beta_2$ by LPS and other inflammatory mediators. Curr Biol 10:974–978

Cheever ML, Sato TK, de Beer T, Kutateladze TG, Emr SD, and Overduin M (2001) Phox domain interaction with PtdIns(3)P targets the Vam7 t-SNARE to vacuole membranes. Nat Cell Biol 3:613–618

Ching TT, Wang DS, Hsu AL, Lu PJ, and Chen CS (1999) Identification of multiple phosphoinositide-specific phospholipases D as new regulatory enzymes for phosphatidylinositol-3,4,5-trisphosphate. J Biol Chem 274:8611–8617

Christoforidis S, Miaczynska M, Ashman K, Wilm M, Zhao L, Yip SC, Waterfield MD, Backer JM, and Zerial M (1999) Phosphatidylinositol-3-OH kinases are Rab5 effectors. Nat Cell Biol 1:249–252

Cockcroft S, and De Matteis MA (2001) Inositol lipids as spatial regulators of membrane traffic. *J Membr Biol* 180:187–194

Collins RF, Schreiber AD, Grinstein S, and Trimble WS (2002) Syntaxins 13 and 7 function at distinct steps during phagocytosis. J Immunol 169:3250–3256

Coppolino MG, Dierckman R, Loijens J, Collins RF, Pouladi M, Jongstra-Bilen J, Schreiber AD, Trimble WS, Anderson R, and Grinstein S (2002) Inhibition of phosphatidylinositol-4-phosphate 5-kinase Iα impairs localized actin remodelling and suppresses phagocytosis. J Biol Chem 277:43849–43857

Cox D, Berg JS, Cammer M, Chinegwundoh JO, Dale BM, Cheney RE, and Greenberg S (2002) Myosin X is a downstream effector of PI(3)K during phagocytosis. Nat Cell Biol 4:469–477

Cox D, Chang P, Zhang Q, Reddy PG, Bokoch GM, and Greenberg S (1997) Requirements for both Rac1 and Cdc42 in membrane ruffling and phagocytosis in leukocytes. J Exp Med 186:1487–1494

Cox D, Dale BM, Kashiwada M, Helgason CD, and Greenberg S (2001) A regulatory role for Src homology 2 domain-containing inositol 5′-phosphatase (SHIP) in phagocytosis mediated by Fcγ receptors and complement receptor 3 ($\alpha_M\beta_2$; CD11b/CD18). J Exp Med 193:61–71

Cox D, Tseng CC, Bjekic G, and Greenberg S (1999) A requirement for phosphatidylinositol 3-kinase in pseudopod extension. J Biol Chem 274:1240–1247

Cullen PJ, Cozier GE, Banting G, and Mellor H (2001) Modular phosphoinositide-binding domains—their role in signalling and membrane trafficking. Curr Biol 11: R882–893

De Corte V, Gettemans J, and Vandekerckhove J (1997) Phosphatidylinositol 4,5-bisphosphate specifically stimulates PP60(c-src) catalyzed phosphorylation of gelsolin and related actin-binding proteins. FEBS Lett 401:191–196

de Renzis S, Sonnichsen B, and Zerial M (2002) Divalent Rab effectors regulate the sub-compartmental organization and sorting of early endosomes. Nat Cell Biol 4:124–133

Della Bianca V, Grzeskowiak M, Dusi S, and Rossi F (1993) Transmembrane signaling pathways involved in phagocytosis and associated activation of NADPH oxidase mediated by FcγRs in human neutrophils. J Leukoc Biol 53:427–438

Della Bianca V, Grzeskowiak M, Lissandrini D, and Rossi F (1991) Source and role of diacylglycerol formed during phagocytosis of opsonized yeast particles and associated respiratory burst in human neutrophils. Biochem Biophys Res Commun 177:948–955

Della Bianca V, Grzeskowiak M, and Rossi F (1990) Studies on molecular regulation of phagocytosis and activation of the NADPH oxidase in neutrophils. IgG- and C3b-mediated ingestion and associated respiratory burst independent of phospholipid turnover and Ca^{2+} transients. J Immunol 144:1411–1417

Deretic V, and Fratti RA (1999) *Mycobacterium tuberculosis* phagosome. Mol Microbiol 31:1603–1609

Desjardins M, Huber LA, Parton RG, and Griffiths G (1994) Biogenesis of phagolysosomes proceeds through a sequential series of interactions with the endocytic apparatus. J Cell Biol 124:677–688

Desjardins M, Nzala NN, Corsini R, and Rondeau C (1997) Maturation of phagosomes is accompanied by changes in their fusion properties and size-selective acquisition of solute materials from endosomes. J Cell Sci 110:2303–2314

Divecha N, Roefs M, Halstead JR, D'Andrea S, Fernandez-Borga M, Oomen L, Saqib KM, Wakelam MJ, and D'Santos C (2000) Interaction of the type Iα PIP kinase with phospholipase D: a role for the local generation of phosphatidylinositol-4,5-bisphosphate in the regulation of PLD2 activity. EMBO J 19:5440–5449

Duclos S, and Desjardins M (2000) Subversion of a young phagosome: the survival strategies of intracellular pathogens. Cell Microbiol 2:365–377

Ebinu JO, Bottorff DA, Chan EY, Stang SL, Dunn RJ, and Stone JC (1998) RasGRP, a Ras guanyl nucleotide-releasing protein with calcium- and diacylglycerol-binding motifs. Science 280:1082–1086

Ellson CD, Anderson KE, Morgan G, Chilvers ER, Lipp P, Stephens LR, and Hawkins PT (2001) Phosphatidylinositol 3-phosphate is generated in phagosomal membranes. Curr Biol 11:1631–1635

Ellson CD, Gobert-Gosse S, Anderson KE, Davidson K, Erdjument-Bromage H, Tempst P, Thuring JW, Cooper MA, Lim ZY, Holmes AB, Gaffney PR, Coadwell J, Chilvers ER, Hawkins PT, and Stephens LR (2001) PtdIns(3)P regulates the neutrophil oxidase complex by binding to the PX domain of p40(phox). Nat Cell Biol 3:679–682

Fallman M, Gullberg M, Hellberg C, and Andersson T (1992) Complement receptor-mediated phagocytosis is associated with accumulation of phosphatidylcholine-derived diglyceride in human neutrophils. Involvement of phospholipase D and direct evidence for a positive feedback signal of protein kinase. J Biol Chem 267:2656–2663

Fallman M, Lew DP, Stendahl O, and Andersson T (1989) Receptor-mediated phagocytosis in human neutrophils is associated with increased formation of inositol phosphates and diacylglycerol. Elevation in cytosolic free calcium and formation of inositol phosphates can be dissociated from accumulation of diacylglycerol. J Clin Invest 84:886–891

Franc NC, White K, and Ezekowitz RA (1999) Phagocytosis and development: back to the future. Curr Opin Immunol 11:47–52

Fratti RA, Backer JM, Gruenberg J, Corvera S, and Deretic V (2001) Role of phosphatidylinositol 3-kinase and Rab5 effectors in phagosomal biogenesis and mycobacterial phagosome maturation arrest. J Cell Biol 154:631–644

Fruman DA, Meyers RE, and Cantley LC (1998) Phosphoinositide kinases. Annu Rev Biochem 67:481–507

Fushman D, Najmabadi-Haske T, Cahill S, Zheng J, LeVine H, 3rd, and Cowburn D (1998) The solution structure and dynamics of the pleckstrin homology domain of G protein-coupled receptor kinase 2 (β-adrenergic receptor kinase 1). A binding partner of G$\beta\gamma$ subunits. *J Biol Chem* 273:2835–2843

Gagnon E, Duclos S, Rondeau C, Chevet E, Cameron PH, Steele-Mortimer O, Paiement J, Bergeron JJ, and Desjardins M (2002) Endoplasmic reticulum-mediated phagocytosis is a mechanism of entry into macrophages. Cell 110:119–131

Gessner JE, Heiken H, Tamm A, and Schmidt RE (1998) The IgG Fc receptor family. Ann Hematol 76:231–248

Ghazizadeh S, Bolen JB, and Fleit HB (1994) Physical and functional association of Src-related protein tyrosine kinases with FcγRII in monocytic THP-1 cells. J Biol Chem 269:8878–8884

Gillooly DJ, Morrow IC, Lindsay M, Gould R, Bryant NJ, Gaullier JM, Parton RG, and Stenmark H (2000) Localization of phosphatidylinositol 3-phosphate in yeast and mammalian cells. EMBO J 19:4577–4588

Godi A, Pertile P, Meyers R, Marra P, Di Tullio G, Iurisci C, Luini A, Corda D, and De Matteis MA (1999) ARF mediates recruitment of PtdIns-4-OH kinase-β and stimulates synthesis of PtdIns(4,5)P$_2$ on the Golgi complex. Nat Cell Biol 1:280–287

Haft CR, de la Luz Sierra M, Barr VA, Haft DH, and Taylor SI (1998) Identification of a family of sorting nexin molecules and characterization of their association with receptors. Mol Cell Biol 18:7278–7287

Han J, Luby-Phelps K, Das B, Shu X, Xia Y, Mosteller RD, Krishna UM, Falck JR, White MA, and Broek D (1998) Role of substrates and products of PI 3-kinase in regulating activation of Rac-related guanosine triphosphatases by Vav. Science 279:558–560

Heiss SG, and Cooper JA (1991) Regulation of CapZ, an actin capping protein of chicken muscle, by anionic phospholipids. Biochemistry 30:8753–8758

Higgs HN, and Pollard TD (2000) Activation by Cdc42 and PIP(2) of Wiskott-Aldrich syndrome protein (WASp) stimulates actin nucleation by Arp2/3 complex. J Cell Biol 150:1311–1320

Hinchliffe KA, Giudici ML, Letcher AJ, and Irvine RF (2002) Type IIα phosphatidylinositol phosphate kinase associates with the plasma membrane via interaction with type I isoforms. Biochem J 363:563–570

Hinkovska-Galcheva V, Boxer LA, Mansfield PJ, Schreiber AD, and Shayman JA (2002) Enhanced phagocytosis through inhibition of de novo ceramide synthesis. J Biol Chem (in press)

Honda A, Nogami M, Yokozeki T, Yamazaki M, Nakamura H, Watanabe H, Kawamoto K, Nakayama K, Morris AJ, Frohman MA, and Kanaho Y (1999) Phosphatidylinositol 4-phosphate 5-kinase α is a downstream effector of the small G protein ARF6 in membrane ruffle formation. Cell 99:521–532

Hughes WE, Cooke FT, and Parker PJ (2000) Sac phosphatase domain proteins. Biochem J 350:337–352

Itoh F, Divecha N, Brocks L, Oomen L, Janssen H, Calafat J, Itoh S, and Dijke Pt P (2002) The FYVE domain in Smad anchor for receptor activation (SARA) is sufficient for localization of SARA in early endosomes and regulates TGF-β/Smad signalling. Genes Cells 7:321–331

Jabril-Cuenod B, Zhang C, Scharenberg AM, Paolini R, Numerof R, Beaven MA, and Kinet JP (1996) Syk-dependent phosphorylation of Shc. A potential link between FcεRI and the Ras/mitogen-activated protein kinase signaling pathway through SOS and Grb2. J Biol Chem 271:16268–16272

Jahraus A, Tjelle TE, Berg T, Habermann A, Storrie B, Ullrich O, and Griffiths G (1998) In vitro fusion of phagosomes with different endocytic organelles from J774 macrophages. J Biol Chem 273:30379–30390

Jankowski A, and Grinstein S (2002) Modulation of the cytosolic and phagosomal pH by the NADPH oxidase. Antioxid Redox Signal 4:61–68

Jenkins GH, Fisette PL, and Anderson RA (1994) Type I phosphatidylinositol 4-phosphate 5-kinase isoforms are specifically stimulated by phosphatidic acid. J Biol Chem 269:11547–11554

Jones DH, Morris JB, Morgan CP, Kondo H, Irvine RF, and Cockcroft S (2000) Type I phosphatidylinositol 4-phosphate 5-kinase directly interacts with ADP-ribosylation factor 1 and is responsible for phosphatidylinositol 4,5-bisphosphate synthesis in the golgi compartment. J Biol Chem 275:13962–13966

Kaplan G (1977) Differences in the mode of phagocytosis with Fc and C3 receptors in macrophages. Scand J Immunol 6:797–807

Karimi K, Gemmill TR, and Lennartz MR (1999) Protein kinase C and a calcium-independent phospholipase are required for IgG-mediated phagocytosis by Mono-Mac-6 cells. J Leukoc Biol 65:854–862

Klarlund JK, Guilherme A, Holik JJ, Virbasius JV, Chawla A, and Czech MP (1997) Signaling by phosphoinositide-3,4,5-trisphosphate through proteins containing pleckstrin and Sec7 homology domains. Science 275:1927–1930

Kunz J, Wilson MP, Kisseleva M, Hurley JH, Majerus PW, and Anderson RA (2000) The activation loop of phosphatidylinositol phosphate kinases determines signaling specificity. Mol Cell 5:1-11

Kurosu H, and Katada T (2001) Association of phosphatidylinositol 3-kinase composed of p110β-catalytic and p85-regulatory subunits with the small GTPase Rab5. J Biochem (Tokyo) 130:73–78

Kusner DJ, Hall CF, and Jackson S (1999) Fcγ receptor-mediated activation of phospholipase D regulates macrophage phagocytosis of IgG-opsonized particles. J Immunol 162:2266–2274

Kutateladze TG, Ogburn KD, Watson WT, de Beer T, Emr SD, Burd CG, and Overduin M (1999) Phosphatidylinositol 3-phosphate recognition by the FYVE domain. Mol Cell 3:805–811

Kwiatkowska K, and Sobota A (1999) Signaling pathways in phagocytosis. Bioessays 21:422–431

Larsen EC, DiGennaro JA, Saito N, Mehta S, Loegering DJ, Mazurkiewicz JE, and Lennartz MR (2000) Differential requirement for classic and novel PKC isoforms in respiratory burst and phagocytosis in RAW 264.7 cells. J Immunol 165:2809–2817

Laux T, Fukami K, Thelen M, Golub T, Frey D, and Caroni P (2000) GAP43, MARCKS, and CAP23 modulate PI(4,5)P$_2$ at plasmalemmal rafts, and regulate cell cortex actin dynamics through a common mechanism. J Cell Biol 149:1455–1472

Lawe DC, Chawla A, Merithew E, Dumas J, Carrington W, Fogarty K, Lifshitz L, Tuft R, Lambright D, and Corvera S (2002) Sequential roles for phosphatidylinositol 3-phosphate and Rab5 in tethering and fusion of early endosomes via their interaction with EEA1. J Biol Chem 277:8611–8617

Lennartz MR (1999) Phospholipases and phagocytosis: the role of phospholipid-derived second messengers in phagocytosis. Int J Biochem Cell Biol 31:415–430

Lorenzi R, Brickell PM, Katz DR, Kinnon C, and Thrasher AJ (2000) Wiskott-Aldrich syndrome protein is necessary for efficient IgG- mediated phagocytosis. Blood 95:2943–2946

Lowry MB, Duchemin AM, Coggeshall KM, Robinson JM, and Anderson CL (1998) Chimeric receptors composed of phosphoinositide 3-kinase domains and Fcγ receptor ligand-binding domains mediate phagocytosis in COS fibroblasts. J Biol Chem 273:24513–24520

Maehama T, and Dixon JE (1998) The tumor suppressor, PTEN/MMAC1, dephosphorylates the lipid second messenger, phosphatidylinositol 3,4,5-trisphosphate. J Biol Chem 273:13375–13378

Marshall JG, Booth JW, Stambolic V, Mak T, Balla T, Schreiber AD, Meyer T, and Grinstein S (2001) Restricted accumulation of phosphatidylinositol 3-kinase products in a plasmalemmal subdomain during Fcγ receptor-mediated phagocytosis. J Cell Biol 153:1369–1380

Massol P, Montcourrier P, Guillemot JC, and Chavrier P (1998) Fc receptor-mediated phagocytosis requires CDC42 and Rac1. EMBO J 17:6219–6229

May RC (2001) Phagocytosis in C. elegans: CED-1 reveals its secrets. Trends Cell Biol 11:150

May RC, Caron E, Hall A, and Machesky LM (2000) Involvement of the Arp2/3 complex in phagocytosis mediated by FcγR or CR3. Nat Cell Biol 2:246–248

May RC, and Machesky LM (2001) Phagocytosis and the actin cytoskeleton. J Cell Sci 114:1061–1077

Mayorga LS, Bertini F, and Stahl PD (1991) Fusion of newly formed phagosomes with endosomes in intact cells and in a cell-free system. J Biol Chem 266:6511–6517

McLaughlin S, Wang J, Gambhir A, and Murray D (2002) PIP_2 and proteins: interactions, organization, and information flow. Annu Rev Biophys Biomol Struct 31:151–175

Murray JT, Panaretou C, Stenmark H, Miaczynska M, and Backer JM (2002) Role of Rab5 in the recruitment of hVps34/p150 to the early endosome. Traffic 3:416–427

Nagaishi K, Adachi R, Matsui S, Yamaguchi T, Kasahara T, and Suzuki K (1999) Herbimycin A inhibits both dephosphorylation and translocation of cofilin induced by opsonized zymosan in macrophagelike U937 cells. J Cell Physiol 180:345–354

Newman SL, Mikus LK, and Tucci MA (1991) Differential requirements for cellular cytoskeleton in human macrophage complement receptor- and Fc receptor-mediated phagocytosis. J Immunol 146:967–974

Nielsen E, Christoforidis S, Uttenweiler-Joseph S, Miaczynska M, Dewitte F, Wilm M, Hoflack B, and Zerial M (2000) Rabenosyn-5, a novel Rab5 effector, is complexed with hVPS45 and recruited to endosomes through a FYVE finger domain. J Cell Biol 151:601–612

Ninomiya N, Hazeki K, Fukui Y, Seya T, Okada T, Hazeki O, and Ui M (1994) Involvement of phosphatidylinositol 3-kinase in Fcγ receptor signaling. J Biol Chem 269:22732–22737

Oancea E, Teruel MN, Quest AF, and Meyer T (1998) Green fluorescent protein (GFP)-tagged cysteine-rich domains from protein kinase C as fluorescent indicators for diacylglycerol signaling in living cells. J Cell Biol 140:485–498

Ofek I, Goldhar J, Keisari Y, and Sharon N (1995) Nonopsonic phagocytosis of microorganisms. Annu Rev Microbiol 49:239-276

Oldham S, Stocker H, Laffargue M, Wittwer F, Wymann M, and Hafen E (2002) The *Drosophila* insulin/IGF receptor controls growth and size by modulating PtdInsP(3) levels. Development 129:4103-4109

O'Shea JJ, Siwik SA, Gaither TA, and Frank MM (1985) Activation of the C3b receptor: effect of diacylglycerols and calcium mobilization. J Immunol 135:3381-3387

Payrastre B, Missy K, Giuriato S, Bodin S, Plantavid M, and Gratacap M (2001) Phosphoinositides: key players in cell signalling, in time and space. Cell Signal 13:377-387

Petiot A, Ogier-Denis E, Blommaart EF, Meijer AJ, and Codogno P (2000) Distinct classes of phosphatidylinositol 3''-kinases are involved in signaling pathways that control macroautophagy in HT-29 cells. J Biol Chem 275:992-998

Prehoda KE, Scott JA, Mullins RD, and Lim WA (2000) Integration of multiple signals through cooperative regulation of the N-WASP-Arp2/3 complex. Science 290:801-806

Raiborg C, Bache KG, Mehlum A, Stang E, and Stenmark H (2001a) Hrs recruits clathrin to early endosomes. EMBO J 20:5008-5021

Raiborg C, Bache KG, Mehlum A, and Stenmark H (2001b). Function of Hrs in endocytic trafficking and signalling. Biochem Soc Trans 29:472-475

Rameh LE, Tolias KF, Duckworth BC, and Cantley LC (1997) A new pathway for synthesis of phosphatidylinositol-4,5-bisphosphate. Nature 390:192-196

Raucher D, Stauffer T, Chen W, Shen K, Guo S, York JD, Sheetz MP, and Meyer T (2000) Phosphatidylinositol 4,5-bisphosphate functions as a second messenger that regulates cytoskeleton-plasma membrane adhesion. Cell 100:221-228

Roggo L, Bernard V, Kovacs AL, Rose AM, Savoy F, Zetka M, Wymann MP, and Muller F (2002) Membrane transport in *Caenorhabditis elegans*: an essential role for VPS34 at the nuclear membrane. EMBO J 21:1673-1683

Rohatgi R, Ho HY, and Kirschner MW (2000) Mechanism of N-WASP activation by CDC42 and phosphatidylinositol-4,5-bisphosphate. J Cell Biol 150:1299-1310

Rupper A, and Cardelli J (2001) Regulation of phagocytosis and endo-phagosomal trafficking pathways in *Dictyostelium discoideum*. Biochim Biophys Acta 1525:205-216

Sakisaka T, Itoh T, Miura K, and Takenawa T (1997) Phosphatidylinositol-4,5-bisphosphate phosphatase regulates the rearrangement of actin filaments. Mol Cell Biol 17:3841-3849

Self AJ, Caron E, Paterson HF, and Hall A (2001) Analysis of R-Ras signalling pathways. J Cell Sci 114:1357-1366

Serrander L, Skarman P, Rasmussen B, Witke W, Lew DP, Krause KH, Stendahl O, and Nusse O (2000) Selective inhibition of IgG-mediated phagocytosis in gelsolin-deficient murine neutrophils. J Immunol 165:2451-2457

Shisheva A (2001) PIKfyve: the road to PtdIns-5-P and PtdIns-3,5-P_2. Cell Biol Int 25:1201-1206

Simonsen A, Lippe R, Christoforidis S, Gaullier JM, Brech A, Callaghan J, Toh BH, Murphy C, Zerial M, and Stenmark H (1998) EEA1 links PI(3)K function to Rab5 regulation of endosome fusion. Nature 394:494-498

Song X, Xu W, Zhang A, Huang G, Liang X, Virbasius JV, Czech MP, and Zhou GW (2001) Phox homology domains specifically bind phosphatidylinositol phosphates. Biochemistry 40:8940–8944

Stauffer TP, Ahn S, and Meyer T (1998) Receptor-induced transient reduction in plasma membrane PtdIns(4,5)P_2 concentration monitored in living cells. Curr Biol 8:343–346

Stenmark H, Aasland R, and Driscoll PC (2002) The phosphatidylinositol-3-phosphate-binding FYVE finger. FEBS Lett 513:77–84

Stephens L, Cooke FT, Walters R, Jackson T, Volinia S, Gout I, Waterfield MD, and Hawkins PT (1994) Characterization of a phosphatidylinositol-specific phosphoinositide 3-kinase from mammalian cells. Curr Biol 4:203–214

Stephens L, Ellson C, and Hawkins P (2002) Roles of PI3Ks in leukocyte chemotaxis and phagocytosis. Curr Opin Cell Biol 14:203–213

Suchard SJ, Hinkovska-Galcheva V, Mansfield PJ, Boxer LA, and Shayman JA (1997) Ceramide inhibits IgG-dependent phagocytosis in human polymorphonuclear leukocytes. Blood 89:2139–2147

Swanson JA, Johnson MT, Beningo K, Post P, Mooseker M, and Araki N (1999) A contractile activity that closes phagosomes in macrophages. J Cell Sci 112:307–316

Takenawa T, and Itoh T (2001) Phosphoinositides, key molecules for regulation of actin cytoskeletal organization and membrane traffic from the plasma membrane. Biochim Biophys Acta 1533:190–206

Terebiznik MR, Vieira OV, Marcus SL, Slade A, Yip CM, Trimble WS, Meyer T, Finlay BB, and Grinstein S (2002) Elimination of host cell PtdIns(4,5)P_2 by bacterial SigD promotes membrane fission during invasion by *Salmonella*. Nat Cell Biol 4:766–773

Tjelle TE, Lovdal T, and Berg T (2000) Phagosome dynamics and function. Bioessays 22:255–263

Tognon CE, Kirk HE, Passmore LA, Whitehead IP, Der CJ, and Kay RJ (1998) Regulation of RasGRP via a phorbol ester-responsive C1 domain. Mol Cell Biol 18:6995–7008

Tolias KF, Hartwig JH, Ishihara H, Shibasaki Y, Cantley LC, and Carpenter CL (2000) Type Iα phosphatidylinositol-4-phosphate 5-kinase mediates Rac-dependent actin assembly. Curr Biol 10:153–156

Vanhaesebroeck B, and Waterfield MD (1999) Signaling by distinct classes of phosphoinositide 3-kinases. Exp Cell Res 253:239–254

Varnai P, Lin X, Lee SB, Tuymetova G, Bondeva T, Spat A, Rhee SG, Hajnoczky G, and Balla T (2002) Inositol lipid binding and membrane localization of isolated pleckstrin homology (PH) domains. Studies on the PH domains of phospholipase Cδ 1 and p130. J Biol Chem 277:27412–27422

Venkateswarlu K, Oatey PB, Tavare JM, and Cullen PJ (1998) Insulin-dependent translocation of ARNO to the plasma membrane of adipocytes requires phosphatidylinositol 3-kinase. Curr Biol 8:463–466

Verkleij AJ, and Post JA (2000) Membrane phospholipid asymmetry and signal transduction. J Membr Biol 178:1-10

Vieira OV, Botelho RJ, and Grinstein S (2002) Phagosome maturation: aging gracefully. Biochem J 366:689–704

Vieira OV, Botelho RJ, Rameh L, Brachmann SM, Matsuo T, Davidson HW, Schreiber A, Backer JM, Cantley LC, and Grinstein S (2001) Distinct roles of class I and class III phosphatidylinositol 3-kinases in phagosome formation and maturation. J Cell Biol 155:19-25

Volinia S, Dhand R, Vanhaesebroeck B, MacDougall LK, Stein R, Zvelebil MJ, Domin J, Panaretou C, and Waterfield MD (1995) A human phosphatidylinositol 3-kinase complex related to the yeast Vps34p-Vps15p protein sorting system. EMBO J 14:3339-3348

Wang J, Gambhir A, Hangyas-Mihalyne G, Murray D, Golebiewska U, and McLaughlin S (2002) Lateral sequestration of phosphatidylinositol-4,5-bisphosphate by the basic effector domain of myristoylated alanine-rich C kinase substrate is due to nonspecific electrostatic interactions. J Biol Chem 277:34401-34412

Witke W, Li W, Kwiatkowski DJ, and Southwick FS (2001) Comparisons of CapG and gelsolin-null macrophages: demonstration of a unique role for CapG in receptor-mediated ruffling, phagocytosis, and vesicle rocketing. J Cell Biol 154:775-784

Wright SD, and Silverstein SC (1983) Receptors for C3b and C3bi promote phagocytosis but not the release of toxic oxygen from human phagocytes. J Exp Med 158:2016-2023

Wurmser AE, and Emr SD (2002) Novel PtdIns(3)P-binding protein Etf1 functions as an effector of the Vps34 PtdIns 3-kinase in autophagy. J Cell Biol 158:761-772

Xu Y, Hortsman H, Seet L, Wong SH, and Hong W (2001) SNX3 regulates endosomal function through its PX-domain-mediated interaction with PtdIns(3)P. Nat Cell Biol 3:658-666

Yin HL, Hartwig JH, Maruyama K, and Stossel TP (1981) Ca^{2+} control of actin filament length. Effects of macrophage gelsolin on actin polymerization. J Biol Chem 256:9693-9697

Yu JW, and Lemmon MA (2001) All phox homology (PX) domains from *Saccharomyces cerevisiae* specifically recognize phosphatidylinositol 3-phosphate. J Biol Chem 276:44179-44184

Zhang Q, Cox D, Tseng CC, Donaldson JG, and Greenberg S (1998) A requirement for ARF6 in Fcγ receptor-mediated phagocytosis in macrophages. J Biol Chem 273:19977-19981

Zhao R, Qi Y, Chen J, and Zhao ZJ (2001) FYVE-DSP2, a FYVE domain-containing dual specificity protein phosphatase that dephosphorylates phosphotidylinositol 3-phosphate. Exp Cell Res 265:329-338

Zheleznyak A, and Brown EJ (1992) Immunoglobulin-mediated phagocytosis by human monocytes requires protein kinase C activation. Evidence for protein kinase C translocation to phagosomes. J Biol Chem 267:12042-12048

Regulation of Endocytosis by Phosphatidylinositol 4,5-Bisphosphate and ENTH Proteins

T. Itoh · T. Takenawa

Department of Biochemistry, Institute of Medical Science, University of Tokyo, 4-6-1 Shirokanedai, Minato-ku, Tokyo, Japan
E-mail: toshiki@ims.u-tokyo.ac.jp

1	Introduction	31
2	PtdIns(4,5)P$_2$ in Endocytosis	33
3	Targets of PtdIns(4,5)P$_2$ in Endocytosis	35
4	ENTH Domains as PtdIns(4,5)P$_2$ Targets Essential for Endocytosis	38
5	Structural Basis of Phosphoinositide–ENTH Interactions	40
6	Conclusions and Perspectives	42
References		43

Abstract Clathrin-mediated endocytosis starts by a recruitment of endocytic proteins to the plasma membrane to induce invagination of lipid bilayer and subsequent vesicle formation. The recruitment of these components requires PtdIns(4,5)P$_2$, a phosphoinositide on the plasma membrane. Although it is well known that the synthesis as well as the disruption of this lipid is important, recent studies have revealed the indispensable roles of direct interaction between PtdIns(4,5)P$_2$ and the endocytic machinery. The ENTH domain is a newly found PtdIns(4,5)P$_2$ binding unit conserved among endocytic proteins like epsins, AP180, and the Hip1/Sla2 family. This review focuses on the essential roles of PtdIns(4,5)P$_2$ and its specific binding partner, the ENTH domain, in clathrin-mediated endocytosis.

1
Introduction

Cells internalize a variety of extracellular materials, from small nutrients to entire cells. This internalization is mediated by a process called endocytosis, in which portions of the plasma membrane invaginate together with extracellular fluids or receptor-ligand complexes and then pinch off to form vesicles containing those materials. In most cases, uptake of relatively small materials by endocytosis (also called pinocytosis) is dependent on a clathrin triskelion (Pearse et al. 2000; Smith and Pearse 1999; Marsh and McMahon 1999). Clathrin-dependent endocytosis is a highly regulated event started by the assembly of the clathrin triskelia resulting in the formation of a clathrin-coated pit. Together with the subsequent invagination and vesicle formation, a series of these events is mediated by various regulatory proteins such as adaptor complex AP2, dynamin, amphiphysin, endophilin, synaptojanin, Eps15, epsin, AP180, and Hip1R and many others (Schmid 1997; Pearse and Robinson 1990; Slepnev and De Camilli 2000; Takei and Haucke 2001). These proteins associate with each other to form an endocytic complex through various domains that recognize and interact with specific peptides. The combination of these kinds of interactions involves the interactions of the SH3 domain with the proline-rich sequence, the EH domain with the NPF motif, the appendage domain with the DPW motif, and the clathrin terminal domain with the clathrin binding motif. Apart from these protein-protein interactions, protein-lipid interactions are also emerging as indispensable requisites that allow endocytosis to start and proceed in a highly regulated manner (Martin 2001; Simonsen et al. 2001; De Camilli et al. 1996). Embedded in the lipid bilayer, phospholipids act as signaling molecules that recruit cytosolic proteins to the plasma membrane and other organelle membranes. Furthermore, such an interaction in itself is the tether force of endocytic proteins that modifies the curvature of the plasma membrane. In particular, phosphatidylinositol polyphosphates (phosphoinositides) are phospholipids with multiple molecular diversities according to the position and number of phosphates on the inositol ring (Simonsen et al. 2001; Itoh and Takenawa 2002). Accordingly, the interaction between a phosphoinositide and its protein target can be modified through phosphorylation and dephosphorylation events. Among all the phosphoinositides present in eukaryotic cells, PtdIns(4,5)P_2 and its binding proteins have been, and continue to be, revealed as indispensable factors in clathrin-mediated endocytosis. For example, most of the endocytic proteins mentioned above are known to interact with, and be regulated by, PtdIns(4,5)P_2. In addition, phosphoinositide-metabolizing enzymes like PtdIns4P 5-kinase and PtdIns(4,5)P_2 5-phosphatase have been reported to be involved in the regulation of endocytosis.

Here, we review the roles of PtdIns(4,5)P$_2$ in the regulatory mechanism of endocytosis and also focus on its specific binding partner, the ENTH domain, which recently emerged as an essential factor in this event.

2
PtdIns(4,5)P$_2$ in Endocytosis

Using perforated cells, Jost et al. succeeded in reconstituting the endocytic event in vitro (Jost et al. 1998). Their assay system was even able to dissect two major steps of tranferrin internalization: clustering of the membrane receptors onto the invaginated coated pits and vesicle formation by membrane fission. In this in vitro assay, the antibiotic neomycin, a strong binding agent of PtdIns(4,5)P$_2$, inhibited the receptor clustering. Furthermore, the PH domain of PLCδ1, which binds specifically to PtdIns(4,5)P$_2$ (Garcia et al. 1995; Lemmon et al. 1995), inhibited both the receptor clustering and vesicle formation steps, whereas the PH domain of Grp1, which is specific to PtdIns(3,4,5)P$_3$ (Klarlund et al. 1997), had no effect. These findings demonstrated that PtdIns(4,5)P$_2$, but not other phosphoinositides like PtdIns(3,4,5)P$_3$, is crucial for all of the sequential steps in clathrin-mediated endocytosis to proceed (Fig. 1A).

Another in vitro experiment revealed the time course of PtdIns(4,5)P$_2$ metabolism during coated vesicle formation (Kinuta et al. 2002). When liposomes were incubated with brain cytosol in vitro, the formation of clathrin-coated vesicles (CCVs) was observed, and even quantified by dynamic light scattering. An increasing amount of PtdIns(4,5)P$_2$ in the liposomes promoted vesicular formation most effectively, accompanied by significant dephosphorylation of the phosphoinositide. In this condition, the dephosphorylation of PtdIns(4,5)P$_2$ was completed within 5 s, whereas vesicular formation took more than 5 min, showing that PtdIns(4,5)P$_2$ turnover through dephosphorylation precedes the formation of vesicles (Fig. 1B).

Dephosphorylation of PtdIns(4,5)P$_2$ on endocytosis is thought to be carried out by a phosphatidylinositol 5-phosphatase, synaptojanin (McPherson et al. 1996). Synaptojanin contains two phosphoinositide phosphatase domains, a Sac domain and a 5-phosphatase domain, that sequentially convert PtdIns(4,5)P$_2$ into PtdIns (Guo et al. 1999). The carboxy-terminal region of a ubiquitously expressed alternatively spliced isoform of synaptojanin (SJ170) binds to endocytic proteins like the

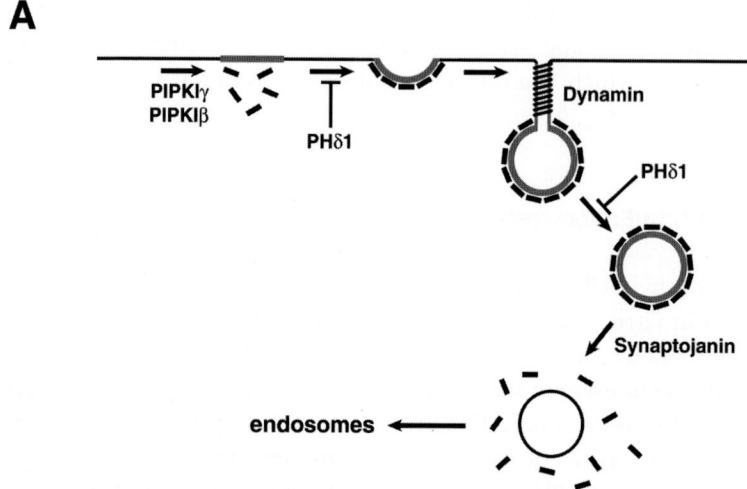

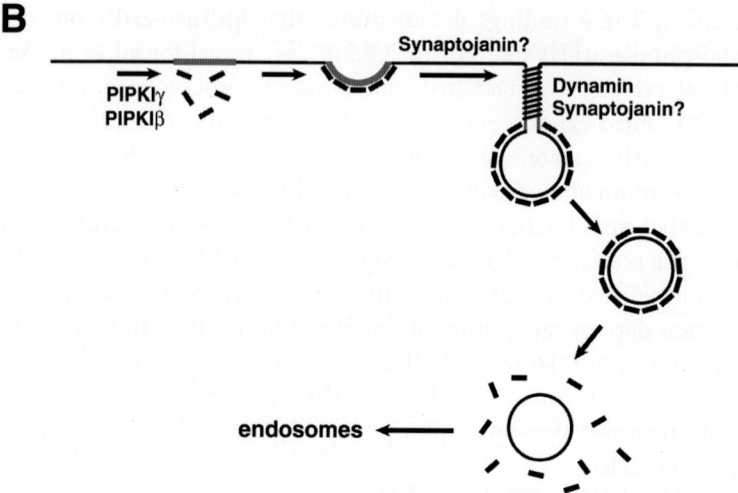

Fig. 1A, B PtdIns(4,5)P$_2$ metabolism in endocytosis. PtdIns(4,5)P$_2$ is synthesized by PIPK Iβ or -Iγ at the plasma membrane, and clathrins are recruited for assembly together with other endocytic proteins. **A** PtdIns(4,5)P$_2$ is required for membrane invagination and subsequent vesicle formation, before it is dephosphorylated by synaptojanin. PH domain of PLC δ1 (*PHδ1*) blocked the vesicle formation, through a competitive inhibition of these protein-lipid interactions. **B** Another study has shown that PtdIns(4,5)P$_2$ dephosphorylation occurs before completion of vesicle formation

clathrin heavy chain, AP2, and Eps15 (Haffner et al. 1997; Haffner et al. 2000). Overexpression of this region perturbed the normal localization of these endocytic proteins as well as transferrin uptake. Thus synaptojanin is one of the components in the endocytic complex that regulates PtdIns(4,5)P$_2$ in the plasma membrane. Gene knockout experiments of synaptojanin in mice (Cremona et al. 1999) and nematodes (Harris et al. 2000) have revealed its indispensable role in vesicle recycling. In both cases, endocytosed synaptic vesicles were observed to accumulate in the nerve terminals in the knockout animals.

The synthesis of PtdIns(4,5)P$_2$ must be the starting signal of vesicular formation that recruits endocytic proteins to the plasma membrane. The kinase responsible for PtdIns(4,5)P$_2$ synthesis, PtdIns4P 5-kinase, has three isoforms (PIPK Iα, β, and γ) (Ishihara et al. 1996; Loijens and Anderson 1996; Ishihara et al. 1998). Among these isoforms, PIPK Iγ is specifically expressed in the brain and is concentrated at synapses. This isoform is localized on the plasma membrane adjacent to clathrin-coated buds, but not on the vesicles themselves, suggesting its role in the initial step of endocytosis, that is, the recruitment of coats to the membrane (Wenk et al. 2001).

The involvement of PIPK in endocytosis has also been reported for another isoform, PIPK Iβ. A truncated fragment of PIPK Iβ was isolated in a screen for genes that restored the signaling ability of an inactive mutant of colony-stimulating factor 1 receptor [CSF-1R (Y809F)] (Davis et al. 1997). This fragment, lacking the first 238 residues [PIPK I$\beta(\Delta1-238)$], seems to act as a dominant-negative form that blocks the internalization of CSF-1R (Y809F) and maintains the receptor on the plasma membrane. Overexpression of PIPK Iβ was demonstrated to enhance the endocytosis of the EGF receptor, whereas a kinase dead mutant inhibited endocytosis by preventing membrane recruitment of the wild-type kinase (Barbieri et al. 2001).

3
Targets of PtdIns(4,5)P$_2$ in Endocytosis

The essential role of PtdIns(4,5)P$_2$ in endocytosis strongly suggested the existence of interactions between the phosphoinositide and endocytic proteins (Table 1). This is actually the case with some endocytic proteins like AP2, β-arrestin, dynamin, and PI3 K C2α.

Table 1 Targets for PtdIns(4,5)P$_2$ in endocytosis. A variety of endocytic proteins interact with PtdIns(4,5)P$_2$ through binding domains (ENTH, PH, BAR) as well as positively charged sequences

Name of protein	Binding site	Specificity	Biological functions
α-Adaptin	Positively charged sequence	PtdIns(3,4,5)P$_3$, PtdIns(4,5)P$_2$	A component of AP2 complex that binds Eps15, epsin, AP180, synaptojanin, amphiphysin, auxilin, etc.
μ2-Adaptin	Positively charged sequence	PtdIns(4,5)P$_2$	A component of AP2 complex that binds receptor proteins via Yxxϕ or D/ExxxLL motif
Epsin	ENTH domain	PtdIns(4,5)P$_2$	Linking between the plasma membrane and endocytic components (AP2, clathrin, Eps15). Formation of clathrin-coated pit
AP180	ENTH domain	PtdIns(4,5)P$_2$	Linking between the plasma membrane and endocytic components (AP2, clathrin). Formation of clathrin-coated pit
Hip1	ENTH domain	Not determined	Linking between actin cytoskeleton (F-actin) and endocytic components (AP2, clathrin). Formation of clathrin-coated pit
Dynamin	PH domain	PtdIns(3,4,5)P$_3$, PtdIns(4,5)P$_2$	Pinching off the invaginated vesicle on GTP hydrolysis by its intrinsic GTPase activity
Amphiphysin	BAR domain	PtdIns(4,5)P$_2$, PtdIns4P	Binds dynamin, synaptojanin, clathrin and AP2
Endophilin	Amino-terminal region (not defined)	Not determined	Binds dynamin, synaptojanin, clathrin and AP2. Containing LPA-acetyltransferase activity

Adaptor complex AP2 is a heterotetramer that consists of α, $\beta 2$, $\mu 2$, and $\sigma 2$-adaptins and acts as a linker between membrane receptors and clathrin (Marsh and McMahon 1999; Robinson and Bonifacino 2001).The α- and μ-adaptins are known to interact with PtdIns(4,5)P$_2$ (Beck et al. 1992; Gaidarov et al. 1996; Rohde et al. 2002). A lysine triad in the amino-terminal region of α-adaptin, as well as three highly conserved lysines in μ-adaptin, were identified as phosphoinositide binding sites (Rohde et al. 2002; Gaidarov and Keen 1999). In both cases, simultaneous substitution of these lysines resulted in abnormal distributions in the cell. Furthermore, overexpression of these mutants blocked normal endocytosis.

After ligand binding, G protein-coupled receptors (GPCRs) are phosphorylated by GPCR kinase and then recognized by β-arrestins to shut off their signals. β-Arrestin binds to clathrin and promotes the internalization of the receptor (Goodman et al. 1996; Goodman et al. 1998). This mode of action is similar to that of AP2, suggesting a role for phosphoinositides in its regulation. Gaidarov et al. demonstrated that β-arrestin binds to PtdIns(3,4,5)P$_3$ and PtdIns(4,5)P$_2$ and three positively charged residues (lysine/arginine) are essential for the binding (Gaidarov et al. 1999). As in the case with AP2, mutation of these residues abolished the binding, and the mutant blocked the internalization of the $\beta 2$-adrenergic receptor in a dominant-negative manner.

Deeply invaginated membranes are pinched off at the final step of vesicle formation. It is well known that this step is mediated by a GTPase called dynamin. Dynamin oligomerizes to form helical tubes on liposomes in vitro and vesiculates the liposomes by GTP hydrolysis (Takei et al. 1995; Sweitzer and Hinshaw 1998). Dynamin has a PH domain that interacts with PtdIns(4,5)P$_2$ as well as PtdIns(3,4,5)P$_3$ (Salim et al. 1996; Barylko et al. 1998). Its GTPase activity is potentiated by phosphoinositide binding (Barylko et al. 1998; Klein et al. 1998). These interactions are also thought to be essential for endocytosis, because a dynamin mutant with a PH domain defective in phosphoinositide binding (K535A/M) blocked endocytosis in a dominant-negative manner (Achiriloaie et al. 1999; Vallis et al. 1999).

PI3K C2$\alpha$is a class II PtdIns 3-kinase that utilizes PtdIns4P and PtdIns(4,5)P$_2$ as substrates to produce PtdIns(3,4)P$_2$ and PtdIns(3,4,5)P$_3$. In vivo, PI3K C2$\alpha$was localized in a punctate structure that overlapped with CCV (Domin et al. 2000; Gaidarov et al. 2001). The amino terminus of PI3K C2αbound to clathrin in vitro, and its kinase activity was markedly potentiated by clathrin binding (Gaidarov et al. 2001). One interesting observation is that PI3K C2α contains a PX domain that exceptionally prefers to bind to PtdIns(4,5)P$_2$ (Song et al. 2001), whereas almost all of the other PX domains bind specifically to PtdIns3P. The role of the PX domain in PI3K C2α is not yet clear, although it may recruit the kinase to areas rich in PtdIns(4,5)P$_2$ where other endocytic proteins are also concentrated. Therefore, PI3K C2α may be activated directly by clathrin to convert PtdIns(4,5)P$_2$ effectively into PtdIns(3,4,5)P$_3$ and thus may rearrange the endocytic complex.

4
ENTH Domains as PtdIns(4,5)P$_2$ Targets Essential for Endocytosis

A protein bound to Eps15 was identified and referred to as epsin (Chen et al. 1998). The central- to carboxy-terminal region of epsin bound simultaneously to three major endocytic proteins, clathrin, AP2, and Eps15. Overexpression of this region blocked endocytosis, indicating its essential role (Chen et al. 1998; Nakashima et al. 1999). In yeast, four homologs of epsin (Ent1p, Ent2p, Ent3p, and Ent4p) have been identified and shown to be required for endocytosis (Wendland et al. 1999). A double-deletion mutant of Ent1p and Ent2p displayed a lethal phenotype that was recovered by introducing a functional version of either gene. With a plasmid shuffling and error-prone PCR technique, it was revealed that the requisite for either gene to be functional was the presence of intact ENTH domains. The ENTH domain had also been found in all epsin family members (epsin 1–3, epsinR/enthorprotin/clint), AP180, CALM, and the other endocytic proteins Hip1 and Hip1R (Kay et al. 1999; Engqvist Goldstein et al. 1999; Mishra et al. 2001; Wasiak et al. 2002; Mills et al. 2003; Hirst et al. 2003; Kalthoff et al. 2002) (Fig. 2). It was also reported that AP180 interacted with InsP$_6$ and that the binding was competed by DiC8PtdIns(3,4,5)P$_3$ or DiC8PtdIns(4,5)P$_2$, suggesting a direct interaction with phosphoinositides (Hao et al. 1997).

Two reports have provided the first evidence that the ENTH domains function as phosphoinositide binding units (Itoh et al. 2001; Ford et al. 2001). A liposome cosedimentation assay revealed that the ENTH domain of epsin1 binds specifically to PtdIns(4,5)P$_2$ (Itoh et al. 2001). The dissociation constant of the binding was calculated to be smaller than 1 μM (0.37 μM), suggesting its high affinity and physiological relevance. The same assay revealed that the ENTH domain of AP180 also bound

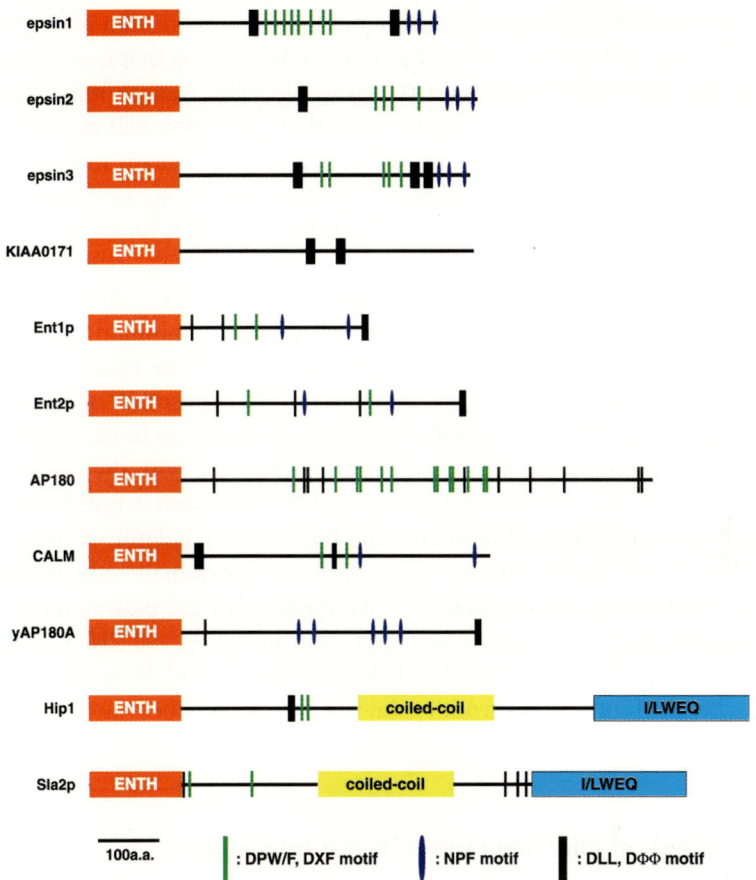

Fig. 2 Endocytic proteins with ENTH domain. Schematic representations for ENTH-containing proteins are shown. DPW/F (DXF) motif is an α-adaptin binding site, whereas DLL (DΦΦ; Φ is hydrophobic amino acid) motif is recognized as a clathrin assembly motif that promotes clathrin-coated pit formation. NPF motif is a binding site for EH domains

specifically to PtdIns(4,5)P$_2$ (Itoh et al. 2001; Ford et al. 2001), confirming the evolutionary conservation of the binding. It was also shown to require several positively charged residues, like Arg63 and Lys76, located in a cleftlike structure of the domain (described in the next section), and substitutions of these residues abolished the binding. Furthermore, internalization of EGF via clathrin-mediated endocytosis was effectively inhibited by overexpression of mutant epsin1 with an ENTH domain rendered incapable of phosphoinositide binding by substitution of Lys76 with Ala (K76A) (Itoh et al. 2001). The effect was equal to that caused by deleting the whole ENTH domain (Chen et al. 1998; Nakashima et al. 1999; Itoh et al. 2001), supporting the idea that phosphoinositide binding is the nature of ENTH domains. Thus the ENTH domain functions as a surface for membrane attachment, allowing epsin to act as an adaptor linking the plasma membrane and clathrin triskelion together with the AP2 complex.

Recently, an additional biochemical function of the epsin ENTH domain on phosphoinositides was reported. When a recombinant protein of full-length epsin was incubated with liposomes from brain lipids (containing phosphoinositides), tubulation of the liposomes was observed (Ford et al. 2002). The region with this activity was in the ENTH domain, as tubulation was also observed by incubation with the recombinant domain alone. This indicates that the ENTH domain is not simply an attachment surface for the membrane but also a regulator of membrane curvature. The tubulation activity of the ENTH domain is believed to be essential for epsin to initiate the invagination of the plasma membrane. Interestingly, this activity was not seen in the AP180 ENTH domain (alternatively called the ANTH domain because of its structural and biochemical differences from epsin, which are described further below).

5
Structural Basis of Phosphoinositide–ENTH Interactions

Both the crystal and NMR structures of the ENTH domains have been solved. The solution structure of the epsin1 ENTH domain revealed that it is composed solely of seven α-helices with a highly labile amino-terminal region(Itoh et al. 2001) (Fig. 3A). On addition of Ins(1,4,5)P_3, the headgroup of PtdIns(4,5)P_2, to the solution, a large chemical shift change was observed in several regions of the epsin1 ENTH domain showing structural changes. In these sites, the essential residues for phosphoinositide binding were identified as Arg63, Trp71, and Lys76 (see the previous section). The most interesting observation in this analysis was the large movement (chemical shift change) of the amino-terminal region, which was highly labile in the absence of Ins(1,4,5)P_3. Deletion of the amino-terminal 18 residues resulted in the failure of phosphoinositide binding. The essential residue in this region was identified

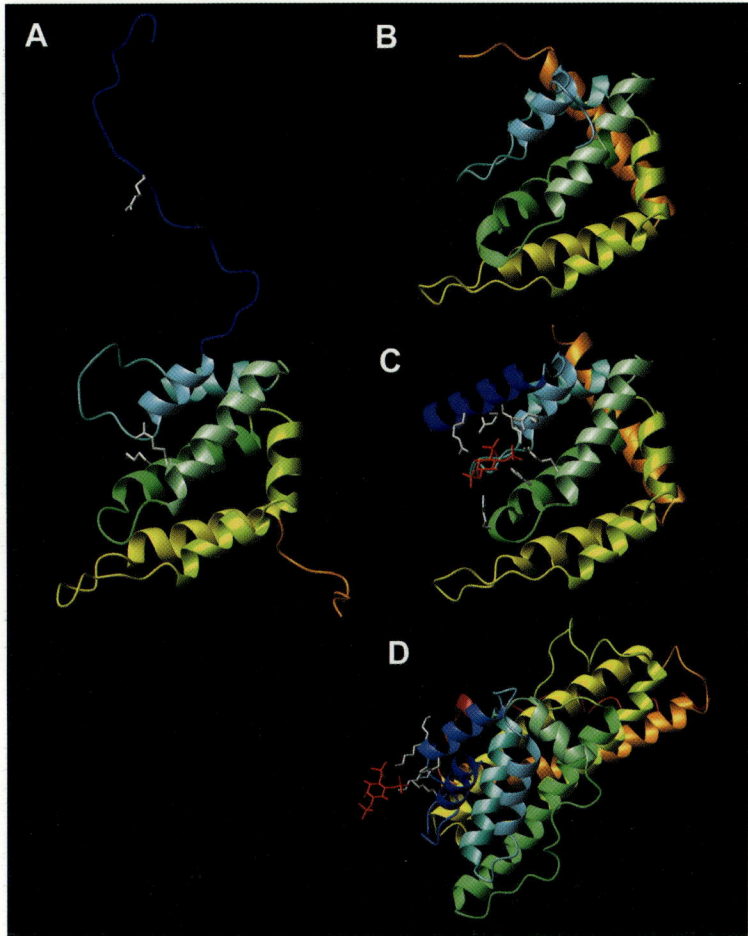

Fig. 3A–D A structural interpretation of the ENTH-PtdIns(4,5)P$_2$ interaction. **A** An NMR structure of epsin ENTH domain (1INZ). In solution, the amino terminal region (*blue*) is unstructured whereas subsequent regions form α-helices. Side chains of Arg8, Arg63 and Lys76 are shown in *white*. **B** A crystal structure of epsin ENTH (1EDU) shows no structural folds for the amino-terminal region in the absence of Ins(1,4,5)P$_3$. **C** In the presence of Ins(1,4,5)P$_3$, the amino-terminal region is observed as an α-helix (*blue*) with side chains surrounding the inositol phosphates (IH0A). **D** A crystal structure of CALM ENTH (ANTH) domain in the presence of Ins(1,4,5)P$_3$ (1HFA). The side chains are facing to the phosphate of Ins(1,4,5)P$_3$ from one direction

as Arg8, as a sole point mutation of Arg8 also abolished the binding. From these results, a structural model for the epsin ENTH domain was presented as follows: PtdIns(4,5)P_2 is positioned in a positively charged cleftlike structure and covered by the amino-terminal region with Arg8 facing toward the phosphates of the phosphoinositide (Itoh et al. 2001).

The importance of the amino-terminal region was confirmed by a recent report describing the tubulation activity of the epsin ENTH domain (Ford et al. 2002). The structure of the epsin ENTH domain cocrystallized with Ins(1,4,5)P_3 was solved and shown to consist of eight α-helices. The additional helix was derived from the amino-terminal region that had been unstructured in the absence of Ins(1,4,5)P_3 (Fig. 3B, C), indicating that it was formed on binding. The helix was amphipathic, with hydrophobic residues lining the outer surface and polar (positively charged) residues facing the inner surface. Of the hydrophobic residues, Leu6 was shown to be involved in the liposome tubulation. This hydrophobicity is thought to be important for the ENTH domain to penetrate into the lipid bilayer to induce membrane curvature.

The crystal structure of the AP180/CALM ENTH domain (also called ANTH or NAP) was also reported to be an all-helix fold consisting of as many as ten α-helices (Ford et al. 2001; Mao et al. 2001) (Fig. 3D). The number of helices simply depended on the definition of the domain in its primary sequence. The AP180/CALM ENTH domain cocrystallized with PtdIns(4,5)P_2 or Ins(4,5)P_2 showed an interaction surface composed of lysines and arginines, and this surface was relatively flat with side chains facing toward a phosphoinositide from one direction, not surrounding it (Fig. 3D). This structure indicated a mode of binding different from that of the epsin ENTH domain, which holds the phosphoinositide in a cleftlike structure capped by an amino-terminal helix (as described above). Together with its inability to cause liposome tubulation (Ford et al. 2002), the AP180/CALM ENTH domain should be categorized as another phosphoinositide binding domain.

6
Conclusions and Perspectives

As discussed in this review, the indispensable roles of PtdIns(4,5)P_2 in endocytosis have been supported by multiple lines of evidence from biochemical studies to genetic approaches. In addition, recent findings about ENTH as a PtdIns(4,5)P_2 binding domain have made, and continue to make, it possible to explain how the phosphoinositide is recognized by endocytic proteins to promote the process of CCV formation. Furthermore, the newly recognized activity of the epsin ENTH domain that induces the formation of membrane curvature is believed to correspond to the very initial step of endocytosis. The important point in this step is the stoichiometry between the ENTH domain and phospholipid on the membrane. It is unclear whether only a single epsin molecule is sufficient for the formation of one entire CCV, or whether multiple epsin molecules localize uniformly on the growing bud to maintain and promote the curvature formation. If the latter possibility is the case, what determines the number of epsin molecules on a single vesicle that eventually defines the size of the CCV itself? Given that epsin is a monomeric protein (Ford et al. 2002) whereas other membrane tubulating proteins like dynamin act in a polymerized form, it is of great interest to discover the determinant of the tubule diameter through controlling the number and distribution of epsin molecules on the membrane.

The synthesis and dephosphorylation of PtdIns(4,5)P_2 in each endocytic process is another issue of great importance. The dephosphorylation that occurs at a very early stage of vesicular formation suggests that not only the recruitment of endocytic proteins by PtdIns(4,5)P_2 but also the (partial or complete) release by PtdIns(4,5)P_2 dephosphorylation is required to complete the entire process of endocytosis. The stepping-stone to examine this hypothesis should be a visualization of phosphoinositide turnover at the site of CCV formation.

For both issues, the most likely breakthrough from the technical aspect might be the visualization of a single phosphoinositide molecule in vivo. More importantly, observation of protein-lipid interactions in vivo at the single molecular level may greatly help us to understand how endocytosis is regulated by PtdIns(4,5)P_2 and its binding proteins.

References

Achiriloaie M, Barylko B and Albanesi JP: Essential role of the dynamin pleckstrin homology domain in receptor-mediated endocytosis. Mol Cell Biol 19:1410–1415,1999

Barbieri MA, Heath CM, Peters EM, Wells A, Davis NJ and Stahl PD: Phosphatidylinositol-4-phosphate-5-kinase-1β is essential for EGF-receptor mediated endocytosis. J Biol Chem 276:47212–47216,2001

Barylko B, Binns D, Lin KM, Atkinson MA, Jameson DM, Yin HL and Albanesi JP: Synergistic activation of dynamin GTPase by Grb2 and phosphoinositides. J Biol Chem 273:3791-3797,1998

Beck KA, Chang M, Brodsky FM and Keen JH: Clathrin assembly protein AP-2 induces aggregation of membrane vesicles: a possible role for AP-2 in endosome formation. J Cell Biol 119:787-796,1992

Chen H, Fre S, Slepnev VI, Capua MR, Takei K, Butler MH, Di Fiore PP and De Camilli P: Epsin is an EH-domain-binding protein implicated in clathrin-mediated endocytosis. Nature 394:793-797,1998

Cremona O, Di Paolo G, Wenk MR, Luthi A, Kim WT, Takei K, Daniell L, Nemoto Y, Shears SB, Flavell RA, McCormick DA and De Camilli P: Essential role of phosphoinositide metabolism in synaptic vesicle recycling. Cell 99:179-188,1999

Davis JN, Rock CO, Cheng M, Watson JB, Ashmun RA, Kirk H, Kay RJ and Roussel MF: Complementation of growth factor receptor-dependent mitogenic signaling by a truncated type I phosphatidylinositol 4-phosphate 5-kinase. Mol Cell Biol 17:7398-7406,1997

De Camilli P, Emr SD, McPherson PS and Novick P: Phosphoinositides as regulators in membrane traffic. Science 271:1533-1539,1996

Domin J, Gaidarov I, SmithME, Keen JH and Waterfield MD: The class II phosphoinositide 3-kinase PI3K-C2alpha is concentrated in the trans-Golgi network and present in clathrin-coated vesicles. J Biol Chem 275:11943-11950,2000

Engqvist Goldstein AE, Kessels MM, Chopra VS, Hayden MR and Drubin DG: An actin-binding protein of the Sla2/Huntingtin interacting protein 1 family is a novel component of clathrin-coated pits and vesicles. J Cell Biol 147:1503-1518,1999

Ford MG, Mills IG, Peter BJ, Vallis Y, Praefcke GJ, Evans PR and McMahon HT: Curvature of clathrin-coated pits driven by epsin. Nature 419:361-366,2002

Ford MG, Pearse BM, Higgins MK, Vallis Y, Owen DJ, Gibson A, Hopkins CR, Evans PR and McMahon HT: Simultaneous binding of PtdIns(4,5)P_2 and clathrin by AP180 in the nucleation of clathrin lattices on membranes. Science 291:1051-1055,2001

Gaidarov I, Chen Q, Falck JR, Reddy KK and Keen JH: A functional phosphatidylinositol 3,4,5-trisphosphate/phosphoinositide binding domain in the clathrin adaptor AP-2 alpha subunit. Implications for the endocytic pathway. J Biol Chem 271:20922-20929,1996

Gaidarov I and Keen JH: Phosphoinositide-AP-2 interactions required for targeting to plasma membrane clathrin-coated pits. J Cell Biol 146:755-764,1999

Gaidarov I, Krupnick JG, Falck JR,Benovic JL and Keen JH: Arrestin function in G protein-coupled receptor endocytosis requires phosphoinositide binding. EMBO J 18:871-881,1999

Gaidarov I, Smith ME, Domin J and Keen JH: The class II phosphoinositide 3-kinase C2alpha is activated by clathrin and regulates clathrin-mediated membrane trafficking. Mol Cell 7:443-449,2001

Garcia P, Gupta R, Shah S, Morris AJ, Rudge SA, Scarlata S, Petrova V, McLaughlin S and Rebecchi MJ: The pleckstrin homology domain of phospholipase C-delta 1 binds with high affinity to phosphatidylinositol 4,5-bisphosphate in bilayer membranes. Biochemistry 34:16228-16234,1995

Goodman OB Jr, Krupnick JG, Santini F, Gurevich VV, Penn RB, Gagnon AW, Keen JH and Benovic JL: Beta-arrestin acts as a clathrin adaptor in endocytosis of the beta2-adrenergic receptor. Nature 383:447–450,1996

Goodman OB Jr, Krupnick JG, Santini F, Gurevich VV, Penn RB, Gagnon AW, Keen JH and Benovic JL: Role of arrestins in G-protein-coupled receptor endocytosis. Adv Pharmacol 42:429–433,1998

Guo S, Stolz LE, Lemrow SM and York JD: SAC1-like domains of yeast SAC1, INP52, and INP53 and of human synaptojanin encode polyphosphoinositide phosphatases. J Biol Chem 274:12990–12995,1999

Haffner C, Di Paolo G, Rosenthal JA and de Camilli P: Direct interaction of the 170 kDa isoform of synaptojanin 1 with clathrin and with the clathrin adaptor AP-2. Curr Biol 10:471–474,2000

Haffner C, Takei K, Chen H, Ringstad N, Hudson A, Butler MH, Salcini AE, Di Fiore PP and De Camilli P: Synaptojanin 1: localization on coated endocytic intermediates in nerve terminals and interaction of its 170 kDa isoform with Eps15. FEBS Lett 419:175–180,1997

Hao W, Tan Z, Prasad K, Reddy KK, Chen J, Prestwich GD, Falck JR, Shears SB and Lafer EM: Regulation of AP-3 function by inositides. Identification of phosphatidylinositol 3,4,5-trisphosphate as a potent ligand. J Biol Chem 272:6393–6398,1997

Harris TW, Hartwieg E, Horvitz HR and Jorgensen EM: Mutations in synaptojanin disrupt synaptic vesicle recycling. J Cell Biol 150:589–600,2000

Hirst J, Motley A, Harasaki K, Peak Chew SY and Robinson MS: EpsinR: and ENTH Domain containing Protein that Interacts with AP-1. Mol Biol Cell 14:625–645,2003

Ishihara H, Shibasaki Y, Kizuki N, Katagiri H, Yazaki Y, Asano T and Oka Y: Cloning of cDNAs encoding two isoforms of 68-kDa type I phosphatidylinositol-4-phosphate 5-kinase. J Biol Chem 271:23611–23614,1996

Ishihara H, Shibasaki Y, Kizuki N, Wada T, Yazaki Y, Asano T and Oka Y: Type I phosphatidylinositol-4-phosphate 5-kinases. Cloning of the third isoform and deletion/substitution analysis of members of this novel lipid kinase family. J Biol Chem 273:8741–8748,1998

Itoh T, Koshiba S, Kigawa T, Kikuchi A, Yokoyama S and Takenawa T: Role of the ENTH domain in phosphatidylinositol-4,5-bisphosphate binding and endocytosis. Science 291:1047–1051,2001

Itoh T and Takenawa T: Phosphoinositide-binding domains. Functional units for temporal and spatial regulation of intracellular signalling. Cell Signal 14:733–743,2002

Jost M, Simpson F, Kavran JM, Lemmon MA and Schmid SL: Phosphatidylinositol-4,5-bisphosphate is required for endocytic coated vesicle formation. Curr Biol 8:1399–1402,1998

Kalthoff C, Groos S, Kohl R, Mahrhold S and Ungewickell EJ: Clint: A novel clathrin-binding ENTH-domain protein at the Golgi. Mol Biol Cell 13:4060–4073,2002

Kay BK, Yamabhai M, Wendland B and Emr SD: Identification of a novel domain shared by putative components of the endocytic and cytoskeletal machinery. Protein Sci 8:435–438,1999

Kinuta M, Yamada H, Abe T, Watanabe M, Li SA, Kamitani A, Yasuda T, Matsukawa T, Kumon H and Takei K: Phosphatidylinositol 4,5-bisphosphate stimulates vesicle formation from liposomes by brain cytosol. Proc Natl Acad Sci USA 99:2842–2847,2002

Klarlund JK, Guilherme A, Holik JJ, Virbasius JV, Chawla A and Czech MP: Signaling by phosphoinositide-3,4,5-trisphosphate through proteins containing pleckstrin and Sec7 homology domains. Science 275:1927–1930,1997

Klein DE, Lee A, Frank DW, Marks MS and Lemmon MA: The pleckstrin homology domains of dynamin isoforms require oligomerization for high affinity phosphoinositide binding. J Biol Chem 273:27725–27733,1998

Lemmon MA, Ferguson KM, O'Brien R, Sigler PB and Schlessinger J: Specific and high-affinity binding of inositol phosphates to an isolated pleckstrin homology domain. Proc Natl Acad Sci USA 92:10472–10476,1995

Loijens JC and Anderson RA: Type I phosphatidylinositol-4-phosphate 5-kinases are distinct members of this novel lipid kinase family. J Biol Chem 271:32937–32943,1996

Mao Y, Chen J, Maynard JA, Zhang B and Quiocho FA: A novel all helix fold of the AP180 amino-terminal domain for phosphoinositide binding and clathrin assembly in synaptic vesicle endocytosis. Cell 104:433–440,2001

Marsh M and McMahonHT: The structural era of endocytosis. Science 285:215–220,1999

Martin TF: PI(4,5)P_2 regulation of surface membrane traffic. Curr Opin Cell Biol 13:493–499,2001

McPherson PS, Garcia EP, Slepnev VI, David C, Zhang X, Grabs D, Sossin WS, Bauerfeind R, Nemoto Y and De Camilli P: A presynaptic inositol-5-phosphatase. Nature 379:353–357,1996

Mills IG, Praefcke GJ, Vallis Y, Peter BJ, Olesen LE, Gallop JL, Butler PJ, Evans PR and McMahon HT: EpsinR: an AP1/clathrin interacting protein involved in vesicle trafficking. J Cell Biol 160:213–222,2003

Mishra SK, Agostinelli NR, Brett TJ, Mizukami I, Ross TS and Traub LM: Clathrin- and AP-2-binding sites in HIP1 uncover a general assembly role for endocytic accessory proteins. J Biol Chem 276:46230–46236,2001

Nakashima S, Morinaka K, Koyama S, Ikeda M, Kishida M, Okawa K, Iwamatsu A, Kishida S and Kikuchi A: Small G protein Ral and its downstream molecules regulate endocytosis of EGF and insulin receptors. EMBO J 18:3629–3642,1999

Pearse BM and Robinson MS: Clathrin, adaptors, and sorting. Annu Rev Cell Biol 6:151–171,1990

Pearse BM, Smith CJ and Owen DJ: Clathrin coat construction in endocytosis. Curr Opin Struct Biol 10:220–228,2000

Robinson MS and Bonifacino JS: Adaptor-related proteins. Curr Opin Cell Biol 13:444–453,2001

Rohde G, Wenzel D and Haucke V: A phosphatidylinositol (4,5)-bisphosphate binding site within mu2-adaptin regulates clathrin-mediated endocytosis. J Cell Biol 158:209–214,2002

Salim K, Bottomley MJ, Querfurth E, Zvelebil MJ,Gout I, Scaife R, Margolis RL, Gigg R, Smith CI, Driscoll PC, Waterfield MD and Panayotou G: Distinct specificity in the recognition of phosphoinositides by the pleckstrin homology domains of dynamin and Bruton's tyrosine kinase. EMBO J 15:6241–6250,1996

Schmid SL: Clathrin-coated vesicle formation and protein sorting: an integrated process. Annu Rev Biochem 66:511–548,1997

Simonsen A, Wurmser AE, Emr SD and Stenmark H: The role of phosphoinositides in membrane transport. Curr Opin Cell Biol 13:485–492,2001

Slepnev VI and De Camilli P: Accessory factors in clathrin-dependent synaptic vesicle endocytosis. Nat Rev Neurosci 1:161–172,2000

Smith CJ and Pearse BM: Clathrin: anatomy of a coat protein. Trends Cell Biol 9:335–338,1999

Song X, Xu W, Zhang A, Huang G, Liang X, Virbasius JV, Czech MP and Zhou GW: Phox homology domains specifically bind phosphatidylinositol phosphates. Biochemistry 40:8940–8944,2001

Sweitzer SM and Hinshaw JE: Dynamin undergoes a GTP-dependent conformational change causing vesiculation. Cell 93:1021–1029,1998

Takei K and Haucke V: Clathrin-mediated endocytosis: membrane factors pull the trigger. Trends Cell Biol 11:385–391,2001

Takei K, McPherson PS, Schmid SL and De Camilli P: Tubular membrane invaginations coated by dynamin rings are induced by GTP-gamma S in nerve terminals. Nature 374:186–190,1995

Vallis Y, Wigge P, Marks B, Evans PR and McMahon HT: Importance of the pleckstrin homology domain of dynamin in clathrin-mediated endocytosis. Curr Biol 9:257–260,1999

Wasiak S, Legendre Guillemin V, Puertollano R, Blondeau F, Girard M, De Heuvel E, Boismenu D, Bell AW, Bonifacino JS and McPherson PS: Enthoprotin: a novel clathrin-associated protein identified through subcellular proteomics. J Cell Biol 158:855–862,2002

Wendland B, Steece KE and Emr SD: Yeast epsins contain an essential N-terminal ENTH domain, bind clathrin and are required for endocytosis. EMBO J 18:4383–4393,1999

Wenk MR, Pellegrini L, Klenchin VA, Di Paolo G, Chang S, Daniell L, Arioka M, Martin TF and De Camilli P: PIP kinase Igamma is the major $PI(4,5)P_2$ synthesizing enzyme at the synapse. Neuron 32:79–88,2001

Membrane Targeting by Pleckstrin Homology Domains

G. E. Cozier · J. Carlton · D. Bouyoucef · P. J. Cullen

Inositide Group, Henry Wellcome Integrated Signaling Laboratories,
Department of Biochemistry, School of Medical Sciences, University of Bristol,
Bristol, BS8 1TD, UK
E-mail: Pete.Cullen@bris.ac.uk

1	Introduction	50
2	PH Domain Classification	52
3	PH Domain Structure	54
4	Mechanisms by Which PII Domains Regulate Phosphoinositide-Mediated Membrane Association	62
4.1	Membrane Association of Proteins Containing PH Domains Which Bind Phosphoinositides with High Affinity and Specificity	63
4.2	Regulated Membrane Association of Proteins with Low-Affinity, Non-specific PH Domains	66
5	PH Domains in the Phosphoinositide-Mediated Allosteric Regulation of Host Proteins	66
6	The Role of PH Domains in the Spatial and Temporal Co-ordination of Phosphoinositide Signalling During Chemotaxis	68
6.1	Recruitment of Effector Proteins to Sites of 3-Phosphoinositide Generation	72
7	PH Domain-Dependent Target to Internal Membranes	74
8	The Potential for PH Domains to Function as Protein Binding Modules	76
9	Conclusions	77
	References	77

Abstract Pleckstrin homology (PH) domains are small modular domains that occur once, or occasionally several times, in a large variety of signalling proteins. In a number of instances, PH domains act to target their host protein to the cytosolic face of cellular membranes through an ability to associate with phosphoinositides. In this review, we discuss recent advances in our understanding of PH domain function. In particular we describe the structural aspects of how PH domains have evolved to bind various phosphoinositides, how PH domains regulate phosphoinositide-mediated association to plasma and internals membranes, and finally raise the issue of PH domains in protein:protein interactions and the allosteric regulation of their host protein.

1
Introduction

The membrane phospholipid phosphatidylinositol is the precursor for a family of lipid second messengers, known collectively as phosphoinositides (Rameh and Cantley 1999; Vanhaesebroeck et al. 2001; Cantrell 2001; Toker 2002). These lipids differ solely in the phosphorylation status of their inositol headgroup (Fig. 1). For the past 50 years, phosphoinositides have attracted considerable attention because of their dual role as precursors of second messenger molecules and as crucial messengers themselves in the localisation and assembly of protein machineries.

For a variety of reasons phosphoinositides are ideally suited to function as spatially restricted membrane second messengers. First, the synthesis and metabolic turnover of phosphoinositides from the relatively abundant phosphatidylinositol precursor can be rapid and highly concentrated within discrete membrane microdomains. Second, the ratio of phosphoinositide to binding partner can be relatively large such that it is possible for a large number of distinct proteins to be targeted to a particular membrane microdomain without saturating the lipid. Third, structurally distinct phosphoinositides can activate distinct downstream effectors. Finally, the potential for rapid, sequential interconversion between phosphorylated forms means that phosphoinositides can confer processivity to membrane signalling events.

A major advance in our understanding of the role of phosphoinositides has stemmed from the identification of a number of highly conserved phosphoinositide-binding modular protein domains. Included with these structurally distinct domains are the epsin amino-terminal homology (ENTH) domain, Fab1, YOTB, Vac1 and EEA1 (FYVE) domain, band 4.1, ezrin, radixin and moesin (FERM) domain, phox homology (PX) domain and the pleckstrin homology (PH) domain (Hurley and Misra 2000; Cullen and Chardin 2000; Hurley and Meyer 2001; Cullen et al. 2001; Itoh and Takenawa 2002). These 'cut and paste' modules

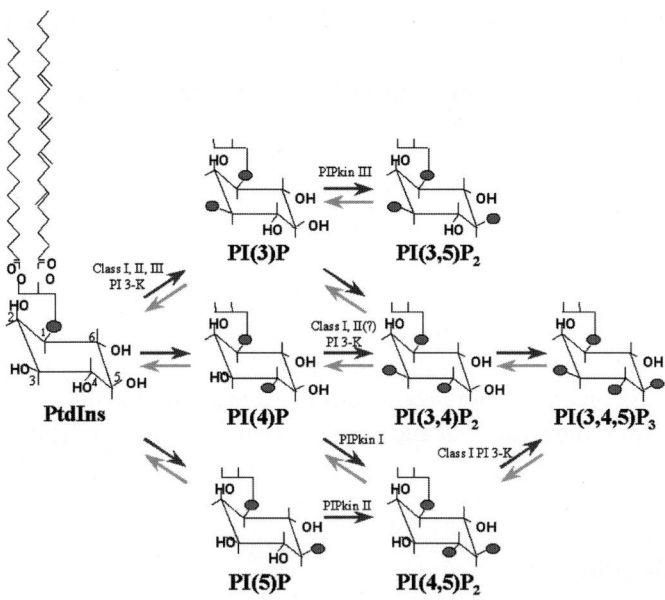

Fig. 1 Phosphoinositide metabolism. Phosphatidylinositol (PtdIns), the basic building block for the intracellular inositol lipids in eukaryotic cells, consists of D-*myo*-inositol-1-phosphate linked via its phosphate group to almost exclusively 1-stearoyl, 2-arachidonoyl diacylglycerol—a fatty acyl composition which may contribute to an adequate packing of the phosphoinositides leading to efficient exposure of their inositol headgroup for interaction with cytosolic proteins. Unlike the headgroups of other phospholipids, the inositol ring of PtdIns can be reversibly phosphorylated at one or a combination of the $3'$, $4'$ or $5'$ positions. PtdIns and its phosphorylated derivatives are collectively referred to as phosphoinositides. PtdIns is the most abundant inositol lipid in mammalian cells under basal conditions, present at levels 10–20 times higher that those of phosphatidylinositol 4-monophosphate [PI(4)P] and phosphatidylinositol 4,5-bisphosphate [PI(4,5)P$_2$], which are present in roughly equal amounts. Of the total singly phosphorylated phosphoinositides in cells, 90%–96% is PI(4)P; phosphatidylinositol 3-monophosphate [PI(3)P] and phosphatidylinositol 5-monophosphate [PI(5)P] each make up about 2%–5%. PI(4,5)P$_2$ is the most abundant of the doubly phosphorylated PIs (>99%); phosphatidylinositol 3,4-bisphosphate [PI(3,4)P$_2$] and phosphatidylinositol 3,5-bisphosphate [PI(3,5)P$_2$] each make up about 0.2% of the PtdInsP$_2$. The levels of phosphatidylinositol 3,4,5-trisphosphate [PI(3,4,5)P$_3$] vary enormously but can be comparable to those of PI(3,4)P$_2$ and PI(3,5)P$_2$

are found in a variety of multidomain proteins. They appear to function by regulating the association of their host protein to a specific subcellular membrane compartment and/or through allosteric modulation.

In the chapter by Birkeland and Stenmark the role of FYVE domain and PX domain-containing proteins in the phosphoinositide-mediated regulation of vesicular traffic is discussed. Here we focus our attention on examining the role of PH domain-containing proteins in phosphoinositide signalling. A number of recent excellent reviews have covered in depth the overall biology of PH domains since their initial identification in 1993 (Lemmon and Ferguson 2000, 2001; Maffucci and Falsca 2001; Lemmon et al. 2002). In the present review we therefore limit ourselves to a discussion of the more recent advances made in this field. We begin by examining the structural and physical characteristics of PH domains.

2
PH Domain Classification

PH domains are small β-sandwich proteins that occur in a range of intracellular proteins including tyrosine kinases [e.g. Bruton's tyrosine kinase (Btk)], serine/threonine kinases [e.g. protein kinase B (PKB)], regulators of small GTP-binding proteins (both GTPase-activating proteins and guanine nucleotide exchange factors), signalling adaptor molecules (e.g. Grb7, IRS-1), proteins that modify membrane phospholipids (phospholipase C isoforms), cytoskeletal proteins (e.g. spectrin) and proteins involved in regulation of membrane traffic (e.g. dynamin) (Fig. 2). Nearly all PH domains so far examined bind membrane phosphoinositides. Binding varies dramatically with respect to phosphoinositide specificity and affinity. PH domains have been classified into two classes on the basis of their phosphoinositide specificity:

Class I These PH domains bind phosphoinositide isomers (and their corresponding inositol phosphates) with high affinity and a high degree of stereospecificity. These PH domains can function independently as phosphoinositide-regulated membrane-association modules. They can be further subdivided into the following:
(a)   PH domains that recognise only $Ins(1,3,4,5)P_4/PtdIns(3,4,5)P_3$ (e.g. Btk) (Fukuda et al. 1996; Kojima et al. 1997).

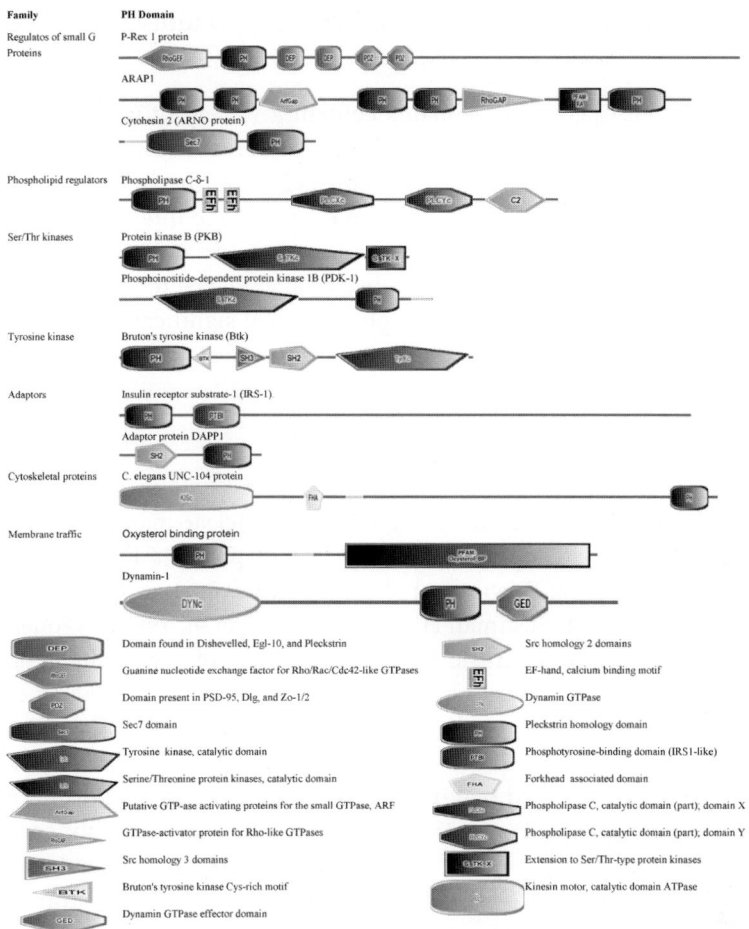

Fig. 2 Schematic representation of the 'cut and paste' nature of pleckstrin homology domains. Selected PH domain-containing proteins that cover a range of physiologic functions are depicted. The other modular domains present within these proteins are also shown

(b) PH domains that bind PtdIns(4,5)P_2 as well as Ins(1,3,4,5)P_4/PtdIns(3,4,5)P_3 (i.e. GAP1^{IP4BP}) (Cozier et al. 2000).

(c) PH domains that prefer Ins(1,4,5)P_3/PtdIns(4,5)P_2 (i.e. PLCδ_1) (Garcia et al. 1995; Lemmon et al. 1995).

(d) PH domains that recognise Ins(1,3,4,5)P_4/PtdIns(3,4,5)P_3 as well as PtdIns(3,4)P_2 (e.g. PKB, DAPP1) (Franke et al. 1997; Frech et al. 1997; Klippel et al. 1997; Anderson et al. 2000).

(e) PH domains that specifically recognise PtdIns(3,4)P$_2$ (i.e. TAPP1) (Dowler et al. 2000).

Class II Although Class I PH domains contain a variety of well-characterised proteins, the vast majority of PH domains display low affinity and a high degree of promiscuity for the ligands mentioned above. Here 'low affinity' means that the interaction with the given phosphoinositide is insufficient to drive membrane localisation. However, there is substantial evidence to suggest that these weak interactions are of physiological significance.

3
PH Domain Structure

PH domain structures have now been solved by nuclear magnetic resonance and X-ray crystallography for a variety of proteins (Yoon et al. 1994; Macias et al. 1994; Downing et al. 1994; Ferguson et al. 1994, 1995, 2000; Timm et al. 1994; Fushman et al. 1995; Zhang et al. 1995; Hyvonen et al. 1995; Hyvonen and Saraste 1997; Zheng et al. 1997; Koshiba et al. 1997; Soisson et al. 1998; Fushman et al. 1998; Baraldi et al. 1999; Dhe-Paganon et al. 1999; Lietzke et al. 2000; Worthylake et al. 2000; Blomberg et al. 2000; Thomas et al. 2001, 2002; Jogl et al. 2002; Rossman et al. 2002). Analysis has revealed that despite their poorly conserved primary sequence, ranging from 7% to 23%, PH domains retain a highly conserved three-dimensional architecture (Fig. 3). The core is a β-sandwich of two nearly orthogonal β-sheets. One sheet consists of four strands (β1 through β4), and the other three (β5–β7). There are six loops connecting the β. Three of these (β1/β2, β3/β4, and β6/β7) are termed the 'variable loops', as they display hypervariable sequences in PH domain alignments. These loops close off one corner of the β-sandwich, whereas an amphipathic carboxy-terminal α-helix closes off the opposite corner. The three variable loops abut the membrane surface when a PH domain binds to a phosphoinositide embedded in a lipid bilayer. Interestingly, the same β-sandwich fold has also been observed in several other domains that share no significant sequence similarities with PH domains (Lemmon and Ferguson 2000). This PH domain 'superfold' includes the Ran-binding domain (Vetter et al. 1999), the Enabled/VASP homology

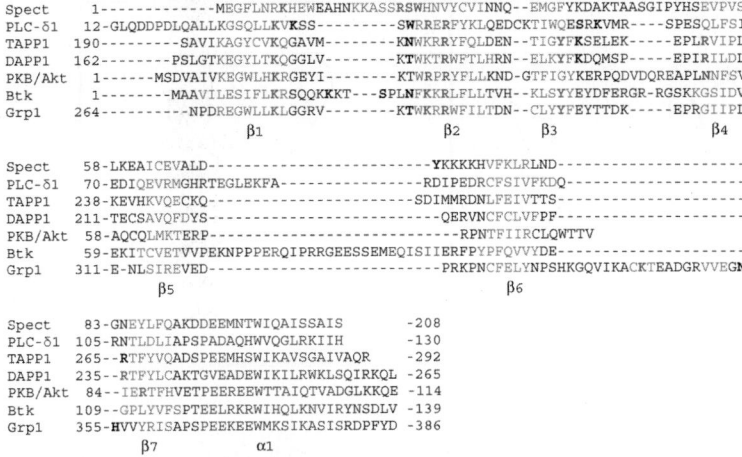

Fig. 3 Structure-based alignment of the sequences of seven PH domains known to bind inositol phosphates with high affinity and/or specificity. The secondary structure is shown for β-strands and α-helices. The major structural features for PH domains are labelled β1–β7 and α1, and residues shown to be involved in binding their respective inositol phosphates are shown in *bold type*

domain-1 (EVH-1) (Prehoda et al. 1999; Federov et al. 1999) and the phosphotyrosine-binding (PTB) domain (Zhou et al. 1995).

To date, the structures of six PH domains have been solved in complex with a bound inositol phosphate headgroup of their cognate phosphoinositide. The PH domains from PLC-δ_1 (Ferguson et al. 1995) and β-spectrin (Hyvonen et al. 1995) have been solved in complex with Ins(1,4,5)P_3, the headgroup of PtdIns(4,5)P_2, and the PH domains from DAPP1 (Ferguson et al. 2000), Btk (Baraldi et al. 1999), Grp1 (Ferguson et al. 2000; Lietzke et al. 2000) and PKB/Akt (Thomas et al. 2002) have all been solved as bound to Ins(1,3,4,5)P_4, the headgroup of PtdIns(3,4,5)P_3. All of these, apart from the PH domain from β-spectrin, have the binding site located in a cleft between the β1/β2-, β3/β4- and β6/β7-loops (Fig. 4), whereas β-spectrin binds Ins(1,4,5)P_3 between the β1/β2- and β5/β6-loops (Fig. 4). The most striking difference between how PLC-δ_1 binds Ins(1,4,5)P_3 and DAPP1, Btk, Grp1 and PKB/Akt all bind Ins(1,3,4,5)P_4 is the orientation of the inositol phosphate. Although there is some variation in position and angle of the Ins(1,3,4,5)P_4 in the binding site of DAPP1, Btk, Grp1 and PKB/Akt [particularly PKB/Akt where Ins(1,3,4,5)P_4 is rotated about 45° and shifted about 4 toward the

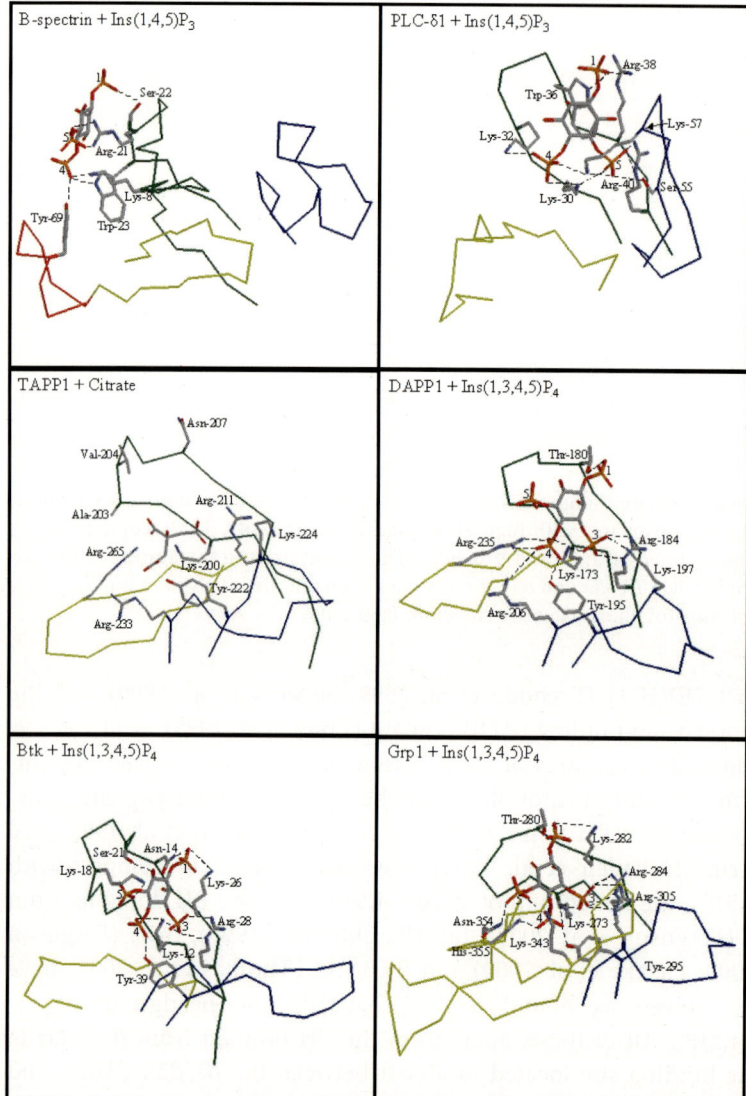

Fig. 4 Structure of six PH domains highlighting the main structural features. The conserved backbone of PH domains is shown as β1–β4 in *dark green*, β5-β7 in *light green*, α1-helix in *dark blue*, other β-strands in *yellow* and other α-helices in *light blue*. The bound inositol phosphates are shown as *sticks*

core of the PH domain], the 3-phosphate is always positioned close to a highly conserved arginine (Arg-184 in DAPP1, Arg-28 in Btk, Arg-284 in Grp1 and Arg-25 in PKB/Akt) (Fig. 5). In comparison, the Ins(1,4,5)P$_3$ in the PH domain of PLC-δ_1 is rotated almost 180° about an axis through the 1- and 4-phosphates, thereby positioning the 5-phosphate near the same arginine (Arg-40 in PLC-δ_1). In all PH domains studied so far that have a high affinity and/or specificity for phosphoinositides and that share the equivalent binding site as observed in PLC-δ_1, DAPP1, Btk, Grp1 and PKB/Akt, this arginine is totally invariant. The difference between how PLC-δ_1 binds Ins(1,4,5)P$_3$ and DAPP1, Btk, Grp1 and PKB/Akt bind Ins(1,3,4,5)P$_4$ gives the first indication that although the PH domains have similar overall structures, and even use some of the same residues in binding inositol phosphates, they have adapted their structures to give different specificities and affinities.

Comparing the binding sites of PLC-δ_1 and β-spectrin, which both bind Ins(1,4,5)P$_3$, it can be seen that although the sites is located in a different part of the overall structure, and therefore use different residues, they are similar in the fact that they provide interactions to all three phosphates. Therefore, they are specific for inositol phosphates that contain this arrangement of phosphates. However, PLC-δ_1 has a much higher affinity than β-spectrin for Ins(1,4,5)P$_3$ [K_D of 210 nM for PLC-δ_1 (Ferguson et al. 1995) compared with 40 mM for β-spectrin (Hyvonen et al. 1995)]. Looking at the two structures of these PH domains binding Ins(1,4,5)P$_3$, it is clear why there is such a difference in affinity, as PLC-δ_1 provides many more residues and interactions with the bound Ins(1,4,5)P$_3$ than can be seen in the β-spectrin structure. There is also a difference in inositol phosphate specificity between these two PH domains. PLC-δ_1 is highly specific for Ins(1,4,5)P$_3$, whereas β-spectrin can also bind Ins(1,3,4,5)P$_4$. Again an examination of the PH domain structures can give an explanation for this difference. β-Spectrin has plenty of space around the 3-hydroxy position to allow a phosphate group to be present. There may not be any additional interactions with this phosphate group, but there is no steric hindrance which would prevent Ins(1,3,4,5)P$_4$ from binding. In contrast PLC-δ_1 has the side chain of lysine-32 located close to the 3-hydroxyl; therefore, there is insufficient space for a phosphate group.

As described above, the PH domains of DAPP1, Btk, Grp1 and PKB/Akt all bind Ins(1,3,4,5)P$_4$ in an equivalent pocket to where PLC-δ_1 binds Ins(1,4,5)P$_3$, albeit in a different orientation. Therefore, there must be

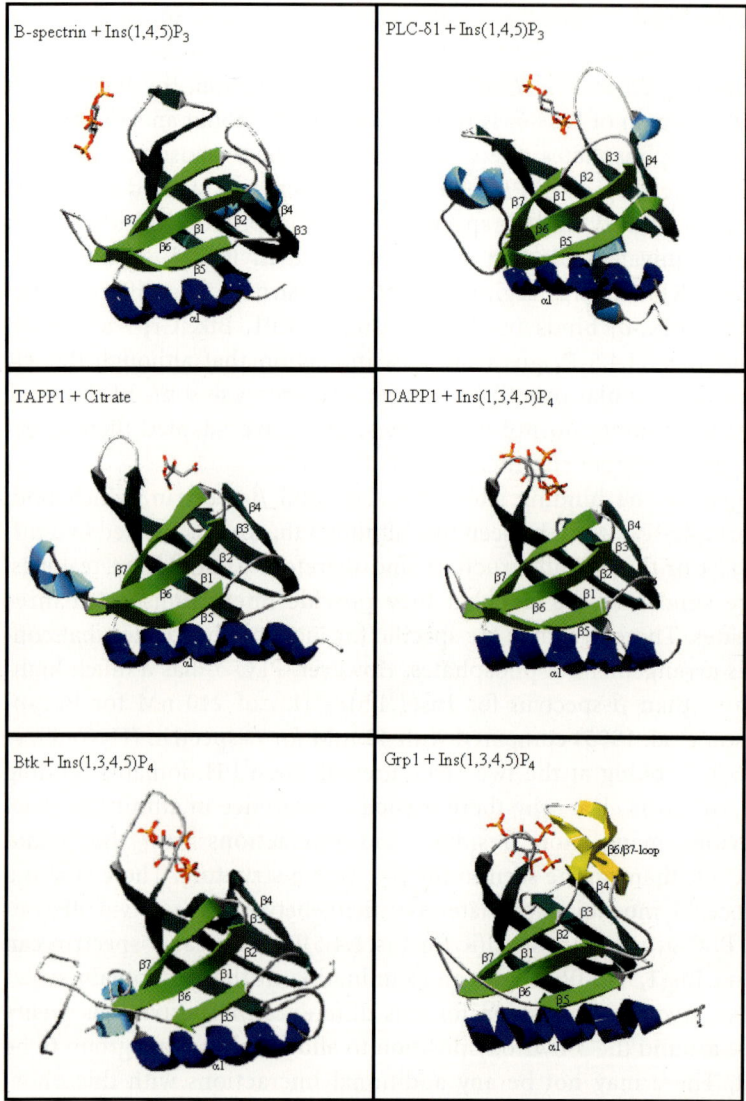

Fig. 5 Schematic representation of PH domains solved in complex with inositol phosphates highlighting the binding sites. The backbone of the loops surrounding the binding sites are shown as: *green-β1/β2, blue-β3/β4, yellow-β6/β7* and also for β-spectrin *red-β5/β6*. The inositol phosphates and side chains are shown as *sticks*, with the phosphates in each inositol phosphate numbered. For clarity only side chains involved in binding the inositol phosphates are shown. The TAPP1 PH domain was only solved in complex with citrate, and this has been shown as a comparison with the other PH domains

differences in the binding sites of these PH domains from the PH domain in PLC-δ_1 which allows space and possible interactions for the extra phosphate. Isakoff et al. (1998) studied the sequences and structures of PH domains that were known to bind 3-phosphoinositides and suggested that the $\beta1/\beta2$-loop was important for binding. They developed a consensus sequence for the $\beta1/\beta2$-loop that would allow binding to 3-phosphoinositides. This consensus had the following characteristics: (a) -2 from the end of the $\beta1$-strand there is a conserved lysine residue; (b) $+2$ residues after this lysine (at the start of the $\beta1/\beta2$-loop) there must be a small residue to allow space for the extra phosphate (glycine, alanine, serine and proline are allowed); (c) the $\beta1/\beta2$-loop must be at least 6 residues in length; (d) there must be at least 1 basic residue on this loop; (e) $+2$ from the start of the $\beta2$-strand there must be a lysine or arginine; and (f) $+4$ from the start of the $\beta2$-strand there must be an arginine. From the PH domains that have been shown to bind 3-phosphoinositides, and those structures solved that are bound to Ins(1,3,4,5)P_4, the majority of these characteristics are conserved. Of all the structures shown in Fig. 5, and the structure of PKB/Akt in complex with Ins(1,3,4,5)P_4, the lysine -2 from the end of the $\beta1$-strand, the arginine at $+4$ of the $\beta2$-strand, the small residue at the start of the $\beta1/\beta2$-loop and the length of the $\beta1/\beta2$-loop all agree with the characteristics suggested by Isakoff et al. (1998). Although the sequences of all the PH domains which bind 3-phosphoinositides described below do agree with the characteristics of having at least one basic residue on the $\beta1/\beta2$-loop, and having a lysine/arginine residue at $+2$ of the $\beta2$-strand, the structures of DAPP1 and PKB/Akt PH domains show that these residues are not actually involved in binding. Instead, as suggested by Ferguson et al. (2000), only those PH domains which are specific for Ins(1,3,4,5)P_4 seem to require the basic residue which can interact with the 5-phosphate, but it can be on either the $\beta1/\beta2$-loop or from another part of the structure such as the $\beta6/\beta7$-loop as found for the PH domain from Grp1. Those PH domains which either bind Ins(1,3,4)P_3, the headgroup of PtdIns(3,4)P_2, or both Ins(1,3,4)P_3 and Ins(1,3,4,5)P_4 may have a basic residue on their $\beta1/\beta2$-loop, but in the examples of DAPP1 and PKB/Akt this residue has no role in binding the 5-phosphate.

To date none of the PH domains which have been solved in complex with Ins(1,3,4,5)P_4 has exactly the same arrangement of residues to give that binding. Below is a summary of each of the PH domain/Ins(1,3,4,5)P_4 complexes to highlight the variety in structure to give the different specificities and affinities. First, DAPP1 binds Ins(1,3,4)P_3 and Ins(1,3,4,5)P_4 with affinities of 0.15 mM and 0.043 mM, respectively. Looking at the structure it can be seen that there are residues located on each of the $\beta1/\beta2$-, $\beta3/\beta4$- and $\beta6/\beta7$-loops which provide interactions with the 1-, 3- and 4-phosphates, with the majority interacting with the 3- and 4-phosphates. From the $\beta1/\beta2$-loop there is lysine-173 (interacts with the 3- and 4-phosphates), threonine-180 (1-phosphate) and arginine-184 (3-phosphate). The $\beta3/\beta4$-loop provides interactions from tyrosine-195 (4-phosphate), lysine-197 (3-phosphate) and arginine-206 (4-phosphate), and on the $\beta6/\beta7$-loop there is arginine-235 (4-phosphate). There are no sidechain interactions with the 5-phosphate, only a few backbone interactions from the $\beta1/\beta2$-loop. This suggests that for DAPP1 the 3- and 4-phosphates are the most important, therefore allowing it to bind both Ins(1,3,4)P_3 and Ins(1,3,4,5)P_4. However, because there are some backbone interactions with the 5-phosphate the affinity for Ins(1,3,4)P_3 is lower than for Ins(1,3,4,5)P_4. The backbone interactions may still form with the 5-hydroxyl in Ins(1,3,4)P_3, but they are likely not to be ideally positioned and therefore weaker.

Also shown in Figs. 4 and 5 is the structure for TAPP1 in complex with citrate. TAPP1 is thought to specifically bind PtdIns(3,4)P_2 (Dowler et al. 2000), or at least have a much lower affinity for PtdIns(3,4,5)P_3 (Ferguson et al. 2000). The sequences of the PH domains of TAPP1 and DAPP1 are quite similar, in particular, all the residues which have been shown to interact with Ins(1,3,4,5)P_4 in DAPP1 are conserved in TAPP1, apart from threonine-180 of DAPP1, which is asparagine-207 in TAPP1 and could therefore have a similar function in binding the 1-phosphate. So it would be predicted that all the interactions from TAPP1 side-chains would, as in DAPP1, be with the 1-, 3- and 4-phosphates. To give the specificity of TAPP1 in only binding PtdIns(3,4)P_2 and not PtdIns(3,4,5)P_3, there must be some reason to block the 5-phosphate position. A comparison of the $\beta1/\beta2$-loops in the structures of DAPP1 and TAPP1 shows that in the case of TAPP1 the loop is significantly closer to the 5-OH position than observed in the DAPP1 structure (Thomas et al. 2001). This change in the loop structure places alanine-203 and valine-204 of TAPP1 in a position that would block a 5-phosphate (Thomas et al. 2001). These residues were individually mutated to glycine residues, with the effect of increasing affinity for PtdIns(3,4,5)P_3, and a double mutation of changing both residues to glycine greatly increased the affinity. Therefore, it appears that in TAPP1 the conformation of the $\beta1/$

β2-loop blocks Ins(1,3,4,5)P_4 binding by placing residues alanine-203 and valine-204 in the space that the 5-phosphate would occupy.

The PH domain from PKB/Akt is similar to DAPP1 in being able to bind both Ins(1,3,4)P_3 and Ins(1,3,4,5)P_4 with similar affinity (Frech et al. 1997). The structure of PKB/Akt was solved in complex with Ins(1,3,4,5)P_4 (Thomas et al. 2002) (NB: at the time of writing the pdb file had not been released; hence the structure is not included in Figs. 4 and 5). Like DAPP1 and TAPP1, PKB/Akt has interactions with the 1-, 3- and 4-phosphates. It has no interactions with the 5-phosphate, explaining why it has similar affinities for both Ins(1,4,5)P_3 and Ins(1,3,4,5)P_4. Again similar to DAPP1, there are interactions from residue side chains located on the β1/β2 (lysine-14, arginine-23 and arginine-25)-, β3/β4 (asparagine-53)- and β6/β7(arginine-86)-loops. However, because Ins(1,3,4,5)P4 in the PKB/Akt binding site is in a slightly different orientation than seen in the other structures, the interactions are a little different. Lysine-14 interacts with both the 3- and 4-phosphates, arginine-23 with both the 1- and 3-phosphates, arginine-25 with the 3-phosphate, asparagine-53 with both the 3- and 4-phosphates and arginine-86 with the 4-phosphate.

Grp1 and Btk both bind Ins(1,3,4,5)P_4 with high specificity and affinity. Therefore, it would be expected that there are interactions to all the 4-phosphates, meaning that all four would have to be present to give maximum binding. By looking at the two structures in Fig. 5 it can be seen that this is indeed the case. In both structures the 3-phosphate interacts with the invariant lysine and arginine (lysine-12 and arginine-28 in Btk; lysine-273 and arginine-284 in Grp1), located towards the end of the β1-strand and +4 from the beginning of the β2-strand, respectively. Both structures also have similar interactions with the 1-phosphate (asparagine-14 and lysine-26 in Btk; threonine-280 and lysine-282 in Grp1), which are from equivalent residues at the end of the β1/β2-loop and +2 from the beginning of the β2-strand. The final interaction conserved in both structures is from a tyrosine towards the end of the β3-strand. Both structures also have additional interactions with the phosphates, but in these interactions there are big differences between the two structures. In Btk the β1/β2-loop plays an important role in providing backbone and side-chain interactions to the 5-phosphate and 6-hydroxyl. Lysine-18 binds the 5-phosphate, and serine-21 interacts with both the 5-phosphate and 6-hydroxyl. The β1/β2-loop in Grp1 is 4 residues shorter than in Btk and does not provide any side chain interactions, but there are backbone interactions with the 5-phosphate. Instead extra side chain interactions come from an extended b6/b7-loop. Residues lysine-273, asparagine-354 and histidine-355 bind to both the 4- and 5-phosphates, and, additionally, arginine-305 from the end of the $\beta3/\beta4$-loop interacts with the 3-phosphate.

A comparison of the PH domains highlights that in one respect they are similar in providing interactions to the phosphates that are important in binding their cognate phosphoinositides. Also, those PH domains which want to specifically block binding of certain phosphoinositides either spread out the interactions to all the phosphates that they want and/or sterically block the position of unwanted phosphates. However, there are large variations in how each PH domain determines its specificity and affinity, the most obvious being the use of a completely different binding site, as observed in β-spectrin, but also changing the orientation of the inositol phosphate in the binding site, with the PH domain from PLC-δ_1 having Ins(1,4,5)P$_3$ rotated almost 180° compared with Ins(1,3,4,5)P$_4$ in the other PH domains and PKB/Akt's PH domain binding Ins(1,3,4,5)P$_4$ in a slightly different orientation. Ignoring β-spectrin, the other PH domains described above have two residues always conserved, that is, the lysine −2 from the end of $\beta1$ and the arginine +4 from the beginning of $\beta2$. No other residues, or even equivalent residues, are conserved in all the PH domains. Instead, each individual PH domain uses its own variation of residues from the $\beta1/\beta2$-, $\beta3/\beta4$- and $\beta6/\beta7$-loops, with each loop having more or less importance depending on the PH domain. Some residues that appear conserved when the sequences of the PH domains are compared are not necessarily conserved in function. Overall, this makes it difficult to predict from the sequence which residues are important and therefore what the specificity of that particular PH domain will be. Following the consensus sequence described by Isakoff et al. (1998), with the modifications suggested by Ferguson et al. (2000), can give a reasonable guideline for predicting specificity.

4
Mechanisms by Which PH Domains Regulate Phosphoinositide-Mediated Membrane Association

Two distinct modes of association appear to be regulated by PH domains. For proteins containing PH domains which bind phosphoinositides with high affinity and specificity, an alteration in the concentration or availability of the phosphoinositide can determine the degree of membrane association, as has been observed for the ARF guanine nucleotide exchange factor ARNO (Venkateswarlu et al. 1998). In contrast, PH domains with low-affinity, non-specific phosphoinositide binding may undergo membrane association independently of any alteration in membrane phosphoinositides. In these cases, cooperative binding with other domains present within the protein or associated proteins may determine membrane binding. In the following section we use specific examples to highlight these distinct modes of binding.

4.1
Membrane Association of Proteins Containing PH Domains Which Bind Phosphoinositides with High Affinity and Specificity

The requirements for receptor-induced recruitment of PH domain-containing proteins to the plasma membrane by binding to the products of PI 3-kinases, namely PtdIns(3,4,5)P_3 and/or PtdIns(3,4)P_2, are that they bind these lipids with high affinity and can detect their presence in membranes that contain substantially higher amounts of other phosphoinositides such as PtdIns(4,5)P_2 and PtdIns(4)P. Several PH domains fulfill these criteria, including those from PKB (Franke et al. 1997; Frech et al. 1997; Klippel et al. 1997), phosphoinositide-dependent protein kinase-1 (PDK-1) (Alessi et al. 1997; Stephens et al. 1998), GAP1^m (Lockyer et al. 1999), Btk (Fukuda et al. 1996; Kojima et al. 1997), the diglycine forms of members of the ARNO/cytohesin family of ARF guanine nucleotide exchange factors (Klarlund et al. 1997, 2000), DAPP1/PHISH (Dowler et al. 1999; Anderson et al. 2000), centaurin-α_1 (Venkateswarlu et al. 1999a), TAPP1 (Dowler et al. 2000), SWAP-70 (Shinohara et al. 2002), LL5β (Paranavitane et al. 2002) and ARAP3 (Arf GAP and Rho GAP with ankyrin repeat and PH domains) (Krugmann et al. 2002). All of these proteins are cytosolic in quiescent cells but become associated with the plasma membrane in a PI 3-kinase-dependent and PH domain-mediated event when cells are treated with growth factors or other agonists (Andjelkovic et al. 1997; Venkateswarlu et al. 1998a, b, 1999, 1999a, b; Anderson et al. 1998, 2000; Lockyer et al. 1999; Watton and Downward 1999; Varnai et al. 1999; Marshall et al. 2002; Shinohara et al. 2002; Krugmann et al. 2002; Paranavitane et al. 2002; Kimber et al. 2002). How the 3-phosphoinositide-induced plasma membrane association leads to the activation of the ability of these proteins to regulate downstream signalling is currently less than clear. One developing theme is that these PH domains function in the recruitment of signalling proteins to specific sites within the plasma membrane which are defined by the elevated level of phosphoinositides. Once recruited to these sites proteins can interact, thereby activating downstream signalling pathways. As an example, the response of PKB to elevation of 3-phosphoinositides results from specific recognition of PtdIns(3,4,5)P_3 and PtdIns(3,4)P_2 by its PH domain (Franke et al. 1997; Frech et al. 1997; Klippel et al. 1997). This results in the plasma membrane recruitment of PKB (Andjelkovic et al. 1997; Watton and Downward 1999), which, although increasing PKB activity directly, allows its subsequent phosphorylation by another kinase, PDK-1. This leads to full PKB activation in vivo. PDK-1 also has a PH domain at its carboxy terminus that binds PtdIns(3,4,5)P_3 (Alessi et al. 1997; Stephens et al. 1998; Anderson et al. 1998). PI 3-kinase products therefore activate PKB principally by recruiting both cytosolic PKB and PDK-1 to the plasma membrane, so that PDK-1 may phosphorylate PKB and activate it completely. This is certainly an oversimplification of the complex regulation of PKB activation (see Vanhaesebroeck and Alessi 2000; Brazil and Hemmings 2001; Toker 2002). However, it serves to highlight that PtdIns(3,4,5)P_3 and PtdIns(3,4)P_2 can activate downstream signalling through their ability to function as a scaffold for the formation of localised plasma membrane signalling complexes.

The role of PH domains in the 3-phosphoinositide-mediated formation of plasma membrane signalling complexes is further highlighted by the PtdIns(3,4,5)P_3 regulation of ARF activation. The ARF family of small GTPases, which includes ARF1–6, are best known for their role in vesicular transport. These proteins function as molecular switches which cycle between a GTP-bound 'on' conformation and a GDP-bound 'off' state. A variety of ARF-specific guanine nucleotide exchange factors (GEFs) activate these GTPases by accelerating the exchange of bound GDP with GTP. Completing the cycle are ARF GTPase-activating proteins (GAPs) that bind to the GTP bound form, stimulating the intrinsic ARF GTPase activity and thereby converting the GTP-bound form into the inactive GDP-bound state. ARF1 localises primarily to the Golgi complex, where it initiates carrier vesicle formation by nucleating the assembly of coat protein complexes. ARF6, on the other hand, has been localised to the plasma membrane/endosomal system, where it regulates the recycling of a subset of membrane proteins (Peters et al. 1995; Radhakrishna et al. 1995). In addition, activation of ARF6 leads to disassembly of actin stress fibres and the formation of structures that resemble lamellipodia (Boshans et al. 2000).

There is convincing evidence that the diglycine forms of the ARNO/cytohesin family of ARF GEFs are effectors of PtdIns(3,4,5)P_3, although the precise role of these proteins in the regulation of ARF proteins is less clear (Jackson and Casanova 2000). They appear to exhibit variable selectivity for the different ARF isoforms in vitro, but there is growing evidence that they may regulate ARF6 in vivo (Santy and Casanova 2001). As with other proteins that contain high-affinity PtdIns(3,4,5)P_3-binding PH domains, these proteins have been shown to be cytosolic in quiescent cells but to undergo a PI 3-kinase- and PH domain-mediated plasma membrane recruitment when cells are treated with growth factors or other agonists (Venkateswarlu et al. 1998a, b, 1999b). Consistent with an important function for PI 3-kinase in the regulation of ARF6 activation has been the identification of ARAP3 [Krugmann et al. 2002; see also the related protein, ARAP1 (Miura et al. 2002)]. This protein has an unusual domain structure, which includes five predicted PH domains and GAP domains for ARF and Rho GTPases (Krugmann et al. 2002). In vitro ARAP3 specifically binds PtdIns(3,4,5)P_3 via its amino-terminal PH domain, and in vivo it undergoes a small but convincing PI 3-kinase- and PH domain-dependent association with the plasma membrane (Krugmann et al. 2002). Interestingly, ARAP3 is a PtdIns(3,4,5)P_3-stimulated ARF6 GAP both in vitro and in vivo (Krugmann et al. 2002). Thus ARAP3 may provide a mechanism whereby PtdIns(3,4,5)P_3 can control not only the GTP loading on ARF6 (via ARNO/cytohesin) but also its GTP hydrolysis (via ARAP3), thereby controlling the complete ARF6 cycle. Thus, as with the interplay between PKB and PDK-1, the PtdIns(3,4,5)P_3- and PH domain-mediated membrane recruitment of these ARF6 regulatory proteins suggests that the spatial dynamics of ARF6 activation and inactivation may occur at specific plasma membrane sites which are defined by the localised elevation in PtdIns(3,4,5)P_3 (Cullen and Venkateswarlu 1999). Given that ARF6 regulates recycling of membrane proteins and the formation of lamellipodia, the PtdIns(3,4,5)P_3-mediated targeting of ARF6 regulatory proteins to specific plasma membranes sites affords a convenient way to alter membrane protein composition and morphology in a localised region of this membrane.

4.2
Regulated Membrane Association of Proteins with Low-Affinity, Non-specific PH Domains

Although 'high-affinity' PH domains function independently as membrane targeting modules, the function of PH domains which bind phosphoinositides weakly and non-specifically is less certain. As discussed above, weak interactions are unlikely by themselves to be sufficient to drive membrane association. However, cooperative interactions with other domains may drive efficient membrane association. For example, the PH domain from β-adrenergic receptor kinase (β-ARK), binds PtdIns(4,5)P$_2$ and G$\beta\gamma$ subunits of heterotrimeric G proteins (Touhara et al. 1994; Fushman et al. 1998). By itself neither interaction can drive membrane association of β-ARK. However, binding of both ligands can drive efficient membrane targeting of β-ARK (Pitcher et al. 1995).

A distinct mode for membrane association driven by weak phosphoinositide-binding affinities may arise through an alteration of the PH domain-containing protein. Here, the PH domain-containing protein may have a sufficiently low affinity for phosphoinositides that the protein remains cytosolic. However, on cellular stimulation, if the protein were to undergo oligomerisation the avidity of the oligomer for phosphoinositides might be sufficiently high to induce membrane association (Lemmon and Ferguson 2000). As an example, the isolated PtdIns(4,5)P$_2$-binding PH domain from dynamin-1 is a cytosolic protein when expressed in mammalian cells (Kavran et al. 1998; Klein et al. 1998). However, because full-length dynamin-1 exists as a tetramer (Muhlberg et al. 1997) altering the oligomerisation state could regulate membrane association of dynamin-1. This is an important property of PH domains which bind weakly to phosphoinositides. If they are present in a protein which can undergo regulated oligomerisation, this could control membrane targeting.

5
PH Domains in the Phosphoinositide-Mediated Allosteric Regulation of Host Proteins

Although the view of the PH domain as a membrane-targeting domain is now generally accepted, it should not be overlooked that phosphoinositide binding may also directly modulate the catalytic activity of the host protein (Drugan et al. 2000; Lowry et al. 2001; Saito et al. 2001). The Dbl-family function as guanine nucleotide exchange factors (GEFs) for the Rho family of small GTPases (Rho, Rac and Cdc42). All Dbl family proteins contain a Dbl homology (DH) domain responsible for GEF activity, which is followed by a PH domain. Although DH-associated PH domains can clearly promote plasma membrane association, the PH domains may also participate directly in the regulation of GEF activity (Han et al. 1998; Ma et al. 1998; Liu et al. 1998; Nimnual et al. 1998; Fleming et al. 2000; Crompton et al. 2000; Russo et al. 2001; Snyder et al. 2001; Welch et al. 2002). Included within this DH/PH group is the recently identified PtdIns(3,4,5)P_3-dependent Rac exchanger (P-Rex1) (Welch et al. 2002). The Rac GEF activity of P-Rex1 is directly, substantially and synergistically activated by PtdIns(3,4,5)P_3 and G$\beta\gamma$ subunits of heterotrimeric G-proteins both in vitro and in vivo (Welch et al. 2002). However, P-Rex1 does not substantially translocate from the cytosol to the plasma membrane sites of PtdIns(3,4,5)P_3 accumulation; rather, the enzyme is partially localised to the membrane in serum-starved cells (Welch et al. 2002). The implication of these results is that PtdIns(3,4,5)P_3 is able to activate the enzyme by inducing a catalytically significant conformational shift or by reorientating P-Rex1 at the plasma membrane surface rather than by simply targeting the protein to the membrane.

How does the PH domain modulate the exchange activity of tandem DH/PH domains? This remains far from clear. Some molecular details have emerged from the crystal structure of the DH/PH fragment from Dbs (Dbl's big sister) in complex with Cdc42 (Rossman et al. 2002). This has revealed that the Dbs PH domain participates with the DH domain in binding Cdc42, primarily through a set of interactions involving the switch 2 region of the GTPase. This interface between Cdc42 and the PH domain is juxtaposed to the phosphoinositide-binding site. It seems plausible, therefore, that the binding of the GTPase and phosphoinositide to the DH and PH domains may not only serve to regulate membrane association but may also induce the proper orientation of the PH domain relative to the DH domain. Interestingly, however, in the crystal structure of the DH/PH fragment of Tiam1 bound to nucleotide-depleted Rac1 (Worthylake et al. 2000), the PH domain fails to engage any part of Rac1. This may be explained by the lack of conservation within the Dbl-family of the residues within the Dbs PH domain which are important for the interaction with Cdc42. Thus only a subset of Dbl-family members may use their PH domains to support nucleotide exchange in the manner described for Dbs (Rossman et al. 2002).

6
The Role of PH Domains in the Spatial and Temporal Co-ordination of Phosphoinositide Signalling During Chemotaxis

The discussion so far has highlighted the role of PH domains in the phosphoinositide-mediated regulation of a number of signalling proteins. However, as is the general case in the post-genomic era, although we know a number of molecular entities whose biological activity is regulated through phosphoinositide binding to PH domains, we have a limited understanding of how phosphoinositides regulate the co-ordinated activation of these proteins to give rise to the desired physiologic effect. Recent work examining the role of phosphoinositides in chemotaxis has, however, begun to address this issue.

Many cells possess the ability to migrate along gradients of chemoattractants. Proper chemotactic function is required for a number of physiologic processes, ranging from axonal guidance and developmental cell migration to neutrophil and leucocyte targeting of invading pathogens, and its function has been implicated in many disease states (Wardlaw et al. 2000; Baggiolini et al. 2001). To achieve controlled chemotaxis, the cell requires an ability to sense both the direction and the intensity of an extracellular stimulus and respond appropriately by co-ordinating its intracellular machinery to undergo directional migration towards the source of the stimulus (reviewed in Comer and Parent 2002). During chemotaxis, the cell adopts a polarised morphology whereby leading and trailing edges are observed to form and traverse the chemotactic gradient. The mass assembly of F-actin at the leading edge (Weiner et al. 1999) suggests that a mechanism exists to preferentially extend the leading edge, thus allowing the motile cell to migrate towards the source of chemoattractant. Mechanisms must also exist to retract the tail end of the cell (Stites et al. 1998; Sanchez-Madrid and del Poso 1999; Chung et al. 2001a). One of the fundamental questions regarding proper chemotactic function is how cells can respond to gradients of chemoattractant that differ by as little as 2% between their posterior and anterior extremes (Parent and Devreotes 1999, Chung et al. 2001b; Katanaev et al. 2001, Weiner et al. 2002a). Mechanisms must exist to amplify the extracellular stimulus and create a polar morphology yielding activation of directional migration. It is becoming clear that localised generation of phosphoinositide gradients has the capacity both to amplify the chemotactic signal and to selectively activate the signalling pathways involved in actin rearrangement leading to cell migration.

A model system for studying chemotaxis is the slime mould *Dictyostelium discoideum*, a unicellular organism that, when starved of nutrients, moves toward secreted adenosine $3',5'$-cyclic monophosphate (cAMP), aggregates and differentiates into spore and stalk cells. The genetic accessibility of this unicellular organism coupled with the role chemotaxis plays in its life cycle makes it an ideal system in which to study chemotactic behaviour. In addition, the study of mammalian neutrophils and leucocytes whose chemotactic migration allows their tracking of pathogens has uncovered addition aspects of the role of phosphoinositides in chemotaxis (Stephens et al. 2002).

The primary event in chemotactic signalling is the binding of chemoattractants to G protein coupled receptors (GPCRs) at the cell surface. The finding that both GPCRs (*D. discoideum* cAMP receptors or mammalian C5a receptors) and their dissociated subunits remain uniformly distributed at the periphery of the cell in chemotaxing *D. dscoideum* and human neutrophils (Jin et al. 2000; Xiao et al. 1997; Servant et al. 1999) suggests that the key polarising stimulus lies downstream of receptor activation and that the signalling pathways mediating actin polymerisation and chemotactic movement must be preferentially activated by these GPCRs at the leading edge of the chemotaxing cell. It is to be noted that in strongly polarised *D. discoideum*, a shallow anterior-to-posterior gradient of G$\beta\gamma$ subunits exists (Jin et al. 2000), indicating that asymmetric distribution of specific G$\beta\gamma$ isoforms could contribute towards generation of the polar signalling responses observed. If, however, the initial receptor activation does not underlie the generation of asymmetric signalling and cytoskeletal alterations required to effect chemotaxis, then what is the next signalling stage that could?

There is a variety of evidence indicating that PI 3-kinase function is important in chemotaxis. Cells treated with PI 3-kinase inhibitors or having a disrupted Class I PI 3-kinase are impaired in chemotaxis (Chung et al. 2001b). Similarly, *D. discoideum* cells bearing a double knockout of two of its 3 PI 3-kinase [PI3K1 and 2 (pi3k1/2null cells)] chemotax poorly, extend multiple pseudopodia in all directions when placed in a chemoattractant gradient and move more slowly than wild-type cells (Funamoto et al. 2002).

It has been shown that the PH domain-containing protein cytosolic regulator of adenylyl cyclase (CRAC) undergoes a PH domain-mediated translocation from the cytoplasm to the leading edge of chemotaxing *D. discoideum* (Parent et al. 1998; Jin et al. 2000). Similar translocations have also been observed for PKB in neutrophils and *D. discoideum* (Servant et al. 2000; Meili et al. 1999) and PhdA (PH domain-containing protein A) in *D. discoideum* (Funamoto et al. 2001). The knowledge that the PH domain of PKB is in effect a biosensor for PtdIns(3,4,5)P_3 and PtdIns(3,4)P_2 (Thomas et al. 2002) suggests that polarised PtdIns(3,4)P_2/PtdIns(3,4,5)P_3 production occurs at the leading edge of chemotaxing cells. Indeed, the chemoattractant-induced translocation of the GFP-tagged PH domain of PKB to the cell's leading edge has been shown to be PI 3-kinase dependent (Meili et al. 1999, 2000). The finding that many proteins implicated in actin rearrangements contain 3-phosphoinositide-binding PH domains suggests that recruitment of these proteins may underly F-actin polymerisation and pseudopod extension at the leading edge, allowing the chemotaxing cell to migrate along a chemotactic gradient.

Recently, using a GFP-tagged PI 3-kinase, PI 3-K1, Funamoto and colleagues (Funamoto et al. 2002), have shown that after chemoattractant stimulation in *D. discoideum* PI 3-K1 can be selectively translocated to the leading edge of the cell, where it is able to generate a localised burst of PtdIns(3,4,5)P_3. The kinetics of localisation are marginally faster than those of the recruitment of PhdA, consistent with the burst of PtdIns(3,4,5)P_3 being required to localise the PhdA and other PH domain-containing proteins. Interestingly, the membrane association of the PI 3-kinase and its activation were separate events; a functional GTP-Ras binding domain is required for PI 3-kinase activation, but not for localisation of the kinase to the leading edge. Indeed, in cells expressing a PI 3-kinase with a mutated Ras binding domain, translocation to the leading edge is observed but the generation of PtdIns(3,4,5)P_3 when exposed to a chemotactic gradient is absent. Whether Ras is preferentially activated at the leading edge of chemotaxing cells or remains uniformly distributed is unclear. These experiments indicate that the gradient of 3-phosphoinositides is in part established by recruitment of a PI 3-kinase to the leading edge of chemotaxing cells, although PI 3-kinase activation requires an upstream cue from GTP-Ras. It is clear, therefore, that generation of PtdIns(3,4,5)P_3 at the leading edge is required for chemotactic migration.

To complement the work on PI 3-kinases, an examination of PTEN (phosphatase and tensin homologue deleted on chromosome 10) activity has shown that it too contributes to the generation of a polar 3-phosphoinositide gradient (Funamoto et al. 2002; Iijima and Devreotes 2002). PTEN is known to convert PtdIns(3,4,5)P_3 to PtdIns(4,5)P_2, thus removing the biologically active 3-phosphoinositide (reviewed in Leslie and Downes 2002). PTEN is predominantly a cytosolic protein but with a visible membrane-bound fraction uniformly distributed around the plasma membrane. On chemoattractant stimulation, PTEN appears to delocalise from the leading edge, effectively allowing the newly synthesised PtdIns(3,4,5)P_3 to build up unabated. In addition, PTEN relocalises to the lateral and trailing edges of the cell, accentuating the PtdIns(3,4,5)P_3 gradient and limiting the localisation of PtdIns(3,4,5)P_3 binding proteins to the leading edge of the cell. The mechanism of PTEN localisation is unclear, although it may be dependent on a putative PtdIns(4,5)P_2 binding motif. It can be envisaged that an increase in PTEN activity at the posterior and lateral edges of the cell will aid generation of its own PtdIns(4,5)P_2 sites, increasing the localisation of PTEN at the lateral and trailing edges.

Iijima and Devreotes (2002) have also shown that in PTEN-null cells, PH domains known to recruit to the leading edge of chemotaxing cells do so but remain associated for up to 10 times longer than in wild-type cells. PTEN-null *D. dscoideum* exhibit less well defined, and occasionally multiple, leading edges and undertake more circuitous chemotactic pathways. Furthermore, they appear to generate pseudopods from the lateral sides of the cell when exposed to a chemotactic gradient. Thus PTEN plays a key role in chemotaxis and acts to suppress spurious pseudopod extension, ensuring the formation of a sole chemotactic axis along which migration can occur. It appears, therefore, that mechanisms exist to convert chemoattractant receptor activation at the front of a cell to localised production of 3-phosphoinositides at the leading edge, furnishing the leading edge with new binding sites for proteins containing PtdIns(3,4)P_2 and PtdIns(3,4,5)P_3 binding PH domains. The finding that PTEN is localised to the trailing edge, possibly creating its own binding sites through localised dephosphorylation of PtdIns(3,4,5)P_3 to PtdIns(4,5)P_2, serves to suppress multiple axis formation and in effect amplify the polarising response. Recruitment of PH domain-containing proteins to the leading edge is predicted to underlie actin polymerisation and pseudopod extension, allowing the cell to migrate along its chemotactic axis, and the regulation of these processes is described below.

6.1
Recruitment of Effector Proteins to Sites of 3-Phosphoinositide Generation

The above evidence indicates that a 3-phosphoinositide gradient is set up on placement of *D. discoideum* and human neutrophil cells in a chemoattractant gradient which serves to recruit PH domain-containing proteins to the leading edge of chemotaxing cells, allowing for pseudopod extension and directed migration. What then are the signalling pathways activated at this leading edge, and how can they couple to actin rearrangements? To extend filopodia or pseudopodia, the chemotaxing cell must enhance actin polymerisation and increase actin nucleation and branching activity selectively at its leading edge (Weiner et al. 1999; Parent and Devreotes 1999; Parent 1999). In human cells, such as the polarised leucocytes or neutrophils, regulation of the actin cytoskeleton is predicted to occur through members of the Rho and ARF family of small GTPases. Mammalian Rho GTPases are subdivided into three classes, RhoA, Rac1 and Cdc42, each of which are predicted to play a distinct role in actin reorganisation. RhoA is involved in formation of stress fibres from actin filaments, Rac is able to organise actin into membranous sheets for generation of membrane ruffles or lamellipodia, and Cdc42 is implicated in the formation of actin-rich filopodia (Hall 1998; Wherlock and Mellor 2002).

As discussed above, the identification of the ARNO/cytohesin family of ARF6 GEFs and the ARF6 GAP ARAP3 as PtdIns(3,4,5)P_3 effectors has highlighted a link between PtdIns(3,4,5)P_3 and the regulation of the small GTPase ARF6. Given that ARF6 regulates recycling of membrane proteins and the formation of lamellipodia, the ability of PtdIns(3,4,5)P_3 to regulate this protein may be of key importance in chemotaxis. Besides ARF6, the ability of PtdIns(3,4,5)P_3 and G$\beta\gamma$ to act synergistically to increase GTP loading on Rac through the regulation of P-Rex1 (Welch et al. 2002), suggests that P-Rex1 may function as a coincidence detector, enhancing GTP loading on Rac in the presence of G$\beta\gamma$ and PtdIns(3,4,5)P_3 (Weiner et al. 2002a). Correlating the requirement for Rac activation in cytoskeletal rearrangements underlying chemotactic behaviour with the finding that G$\beta\gamma$ subunits remain uniformly distributed around the plasma membrane in chemotaxing neutrophils (Jin et al. 2000, Xiao et al. 1997, Servant et al. 1999), P-Rex1 is ideally placed to co-ordinate Rac activation in response to a polarised PtdIns(3,4,5)P$_3$ burst in chemotaxing cells.

An additional protein capable of regulating actin dynamics at the leading edge is *D. discoideum* PhdA, one of the first proteins observed to translocate to the leading edge of chemotaxing cells. PhdA is implicated in the proper assembly of F-actin at the leading edge in response to chemoattractant stimulation (Funamoto et al. 2001; Parent et al. 1998). PhdA-null *D. discoideum* chemotax poorly in response to cAMP stimulation and exhibit polarisation defects and ,in agreement with this, exhibit reduced levels of F-actin polymerisation and show delayed responses in extending new pseudopods on altering the direction of chemoattractant stimulation. Additionally, mutations to the PhdA PH domain to abrogate PtdIns(3,4,5)P$_3$ binding activity confer identical impairments in actin rearrangements. It has been suggested that PhdA acts as a scaffold, recruited to the leading edge of chemotaxing *D. discoideum* cells, and serves to localise other proteins involved in actin rearrangements to the leading edge (Funamoto et al. 2001; Lemmon et al. 2002).

Of the other proteins known to translocate to the leading edge of chemotaxing cells, CRAC is known only to activate adenylyl cyclase (Funamoto et al. 2001) though the role of adenylyl cyclase activation in regulation of the actin cytoskeleton is currently unclear. Akt/PKB, also observed to translocate to the leading edge in both *D. discoideum* and leucocytes, has been shown, like PhdA, to be essential for proper chemotaxis (Meili et al. 1999). PKB activation is impaired in pi3k1/2-null *D. discoideum* and PI 3-kinase γ-null neutrophils and PKB-null cells chemotax poorly (Meili et al. 1999, 2000). As yet, a clear pathway downstream of PKB is unclear; nonetheless, PKB appears essential in co-ordinating the chemotactic response. Although it is recruited to the leading edge, there is good evidence to suggest that it plays an important role at the other end of the cell.

So we have, for motile cells, an ability to evoke chemotaxis through the generation of highly polarised 3-phosphoinositide gradients which act to recruit a number of PH domain-containing effector proteins implicated in the formation of dynamic actin structures at the front of the cell. A key feature of this chemotactic behaviour is the role of recruitment of PH domain-containing proteins to this steep phosphoinositide gradient. Without their ability to function as receptors for transiently generated 3-phosphoinositides, the exquisite spatiotemporal activation of signalling pathways mediating cytoskeletal rearrangements would be lost.

7
PH Domain-Dependent Target to Internal Membranes

So far we have highlighted the importance of PH domains in the phosphoinositide-mediated targeting of signaling proteins to the plasma membrane. Given that phosphoinositides are also present on a variety of internal membranes (Cockcroft and De Matteis 2001; Toker 2002), an important issue is whether PH domains may also target proteins to other organelles, such as endosomes and the Golgi complex. Although no data exist in favour of endosomal association, evidence is emerging for a role of PH domains in Golgi targeting.

It is clear that the Golgi complex is an active and highly controlled phosphoinositide-metabolising centre and that phosphoinositides play an important role in traffic, especially at the *trans*-Golgi network (TGN) (for an excellent detailed discussion of phosphoinositides and the Golgi complex, see De Matteis et al. 2002). PtdIns(4,5)P_2 has been proposed to play a role in endoplasmic reticulum-to-Golgi transport (Godi et al. 1998), in the formation/release of post-Golgi transport intermediates and in maintaining the structural integrity and function of the Golgi complex (Siddhanta et al. 2000, 2002; Sweeney et al. 2002; Watt et al. 2002). In addition, the level of phosphatidylinositol 4-monophosphate [PtdIns(4)P] appears to be important in controlling exit from the Golgi complex (Walch-Solimena and Novick 1999; Hama et al. 1999; Audhya et al. 2000). The ability of PtdIns(4)P to control exit from the Golgi complex in multiple directions suggests that this phosphoinositide may be required for elementary processes which are likely to be common to different trafficking pathways, such as budding and/or fusion processes. How do these phosphoinositides exert their effects on Golgi function?

PtdIns(4,5)P_2 has been proposed to function by controlling the spectrin and actin machineries, both of which play a role in the organisation and function of the Golgi complex (De Matteis and Morrow 1998; Lorra and Huttner 1999). Although spectrin does contain a PtdIns(4,5)P_2-binding PH domain, it is not sufficient to target spectrin to the Golgi complex. Efficient Golgi-targeting requires the amino terminus of the protein along with the carboxy-terminal PH domain (De Matteis and

Morrow 2000). A potential Golgi effector for PtdIns(4)P is the PH domain-containing oxysterol-binding protein (OSBP) (Levine and Munro 1998; Levine and Munro 2001; Li et al. 2002; Levine and Munro 2002; Wyles et al. 2002). This is a member of a family of eukaryotic proteins that bind oxygenated derivatives of cholesterol and may regulate, directly or indirectly, sterol synthesis and other cell functions, including membrane traffic at the Golgi complex (Xu et al. 2001). In vitro the PH domain of OSBP binds PtdIns(4)P and PtdIns(4,5)P$_2$ (Levine and Munro 1998), although in vivo it is the ability of this domain to bind PtdIns(4)P that contributes to Golgi targeting (Levine and Munro 2002). As with spectrin, however, phosphoinositide binding alone appears insufficient for efficient Golgi targeting (Levine and Munro 2002). A second, PtdIns(4)P-independent, factor is required which is dependent on the activity of the Golgi-specific GTPase Arf1p (Levine and Munro 2002). Thus a model is emerging of multiple binding sites in which phosphoinositide binding is only one of the determinants which make up the final affinity of binding that is required for efficient targeting of PH domain-containing proteins to the Golgi complex.

Recently, several proteins have been identified with PH domains which are related to that of OSBP, although they lack oxysterol-binding domains and show no other similarity to OSBP. These include Goodpasture antigen binding protein (GPBP), a human protein which has homologues in invertebrates but not yeast (Raya et al. 1999), and PtdIns(4)P adaptor protein-1 (FAPP-1) a protein so far only found in mammals (Dowler et al. 2000). Indeed, it was recently reported that PH^{GPBP} and PH^{FAPP-1} when expressed in COS cells co-localise with the Golgi marker TGN46 (Levine and Munro 2002). In both cases efficient Golgi targeting requires PtdIns(4)P production in vivo (Levine and Munro 2002). Thus the Golgi-recognising PH domains are found not only in OSBP and its relatives but also in GPBP and FAPP-1, members of two additional protein families. The functions of these proteins are unknown, but both GPBP and a FAPP-1 relative, FAPP-2, contain additional domains related to known lipid binding proteins [StAR and glycolipid transfer protein, respectively (Abe 1990; Tsujishita and Hurley 2000)], suggesting that all three protein families could be involved in the metabolism or sorting of lipids in the Golgi.

8
The Potential for PH Domains to Function as Protein Binding Modules

The major emphasis of our discussion has centered on the premise that PH domains function as binding sites for phosphoinositides. It should be remembered, however, that PH domains were originally suggested to function as protein binding domains. Although this possibility is not currently in vogue, evidence continues to emerge that for certain PH domains their ability to bind proteins may be important in the regulation of their host protein. For example, the PH domains from β-ARK and Btk bind Gβγ subunits of heterotrimeric G proteins (Touhara et al. 1994; Tsukuda et al. 1994; Wang et al. 1994, 1995; Fushman et al. 1998) and the PH domain of Btk, PKB and protein kinase D has been reported to bind various PKC isoforms (Yao et al. 1994, 1997; Konishi et al. 1994, 1995; Waldron et al. 1999). In none of these cases, however, has a clear physiologic relevance for these interactions been demonstrated. More recently, PH domains have been proposed to interact with RACK1, the receptor for activated PKC (Rodriguez et al. 1999; Koehler and Moran 2001), the α from the G_{12} family of heterotrimeric G-proteins (Jiang et al. 1998), a protein termed BAP-135 (Yang and Desiderio 1997) and filamentous actin (Yao et al. 1999). Furthermore, the PH domains from Trio bind filamin (Bellanger et al. 2000), the PH domain of Etk has been described to interact with the FERM domain of focal adhesion kinase (FAK) (Chen et al. 2001) and the PH domain of insulin receptor substrate-1 (IRS-1) binds to proteins containing acidic amino acid residues, such as nucleolin (Burks et al. 1998) and PHIP (PH-interacting protein) (Farhang-Fallah et al. 2000).

Over all, therefore, many different protein ligands have been suggested or reported to bind PH domains. At present, however, there are no candidates for consideration as a general protein target for PH domains. Nevertheless, evidence for protein-protein interactions driven by PH domains is growing. Considering that cooperative protein-protein interactions coupled with phosphoinositide-binding may together regulate membrane association, protein-protein interaction could have a profound effect on the regulation of PH domain-containing proteins.

9
Conclusions

Phosphoinositide binding to PH domains is generally of low affinity and rapidly reversible. Such characteristics have allowed for a highly plastic phosphoinositide signalling system, in which proteins choose between a series of random diffusions in the cytosol, and phosphoinositide-mediated membrane association/dissociation events. Thus PH domain-containing proteins are constantly sampling the cytosolic face of membranes seeking out the presence of their cognate phosphoinositide (Teruel and Meyer 2000). Elevation of the partner phosphoinositide, after cellular stimulation, drives the local enrichment or association of the PH domain-containing signalling protein. As individual phosphoinositides can be readily produced and degraded, half-lives of phosphoinositide-signalling protein complexes can be extremely brief. PH domain-containing proteins should therefore be considered as restless proteins which travel throughout the cytosol, seeking out their cognate phosphoinositide (Teruel and Meyer 2000).

Acknowledgements. We thank the other members of the Inositide Group for helpful discussion and proofreading of the manuscript. The Medical Research Council and the Biotechnology and Biological Sciences Research Council fund work in the authors' laboratory. PJC is a Lister Institute Research Fellow.

References

Abe, A. (1990) Primary structure of glycolipid transfer protein from pig brain. J. Biol. Chem. 265, 9634–9637

Alessi, D.R., James, S.R., Downes, C.P., Holmes, A.B., Gaffney, P.R., Reese, C.B. and Cohen, P. (1997) Characterisation of a 3-phosphoinositide-dependent protein kinase which phosphorylates and activates protein kinase Bα. Curr. Biol. 7, 261–269

Anderson, K.E., Coadwell, J.W., Stephens, L.R. and Hawkins, P.T. (1998) Translocation of PDK-1 to the plasma membrane is important in allowing PDK-1 to activate protein kinase B. Curr. Biol. 8, 684–691

Anderson, K.E., Lipp, P., Bootman, M.D., Ridley, S.H., Coadwell, J., Ronnstrand, L., Lennartsson, J., Holmes, A.B., Painter, G.F., Thuring, J., Lim, Z.-Y., Erdjument-Bromage, H., Grewal, A., Tempst, P., Stephens, L.R. and Hawkins, P.T. (2000) DAPP1 undergoes a PI 3-kinase-dependent cycle of plasma membrane recruitment and endocytosis upon cell stimulation. Curr. Biol. 10, 1403–1412

Andjelkovic, K., Alessi, D.R., Meier, R., Fernandez, A., Lamb, N.J., Frech, M., Cron, P., Cohen, P., Lucocq, J.M. and Hemmings, B.A. (1997) Role of translocation in the activation and function of protein kinase B. J. Biol. Chem. 272, 31515–31524

Audhya, A., Foti, M. and Emr, S.D. (2000) Distinct role for the yeast phosphatidylinositol 4-kinases, Stt4p and Pik1p, in secretion, cell growth, and organelle membrane dynamics. Mol. Biol. Cell 11, 2673–2689

Baggiolini, M. (2001) Chemokines in pathology and medicine. J. Intern. Med. 250, 91–104

Baraldi, E., Carugo, K.D., Hyvonen, M., Surdo, P.L., Riley, A.M., Potter, B.V.L., O'Brien, R., Ladbury, J.E. and Saraste, M. (1999) Structure of the PH domain from Bruton's tyrosine kinase in complex with inositol 1,3,4,5-tetrakisphosphate. Structure 7, 449–460

Bellanger, J.M., Astier, C., Sardet, C., Ohta, Y., Stossel, T.P. and Debant, A. (2000) The Rac1- and RhoG-specific guanine nucleotide exchange factor domain of Trio targets filamin to remodel cytoskeletal actin. Nat. Cell Biol. 2, 888–892

Blomberg, N., Baraldi, E., Sattler, M., Saraste, M. and Nilges, M. (2000) Structure of the pleckstrin homology domain from the *C. elegans* muscle protein UNC-89 suggests a novel function. Structure 8, 1079–1087

Boshans, R.L., Szanto, S., van Aelst, L. and D'Souza-Schorey, C. (2000) ADP-ribosylation factor 6 regulates actin cytoskeleton remodelling in coordination with Rac1 and RhoA. Mol. Cell. Biol. 20, 3685–3694

Brazil, D.P. and Hemmings, B.A. (2001) Ten years of protein kinase B signalling: a hard Akt to follow. Trends Biochem. Sci. 26, 657–664

Burks, D.J., Wang, J., Towery, H., Ishibashi, O., Lowe, D., Riedel, H. and White, M.F. (1998) IRS pleckstrin homology domains bind to acidic motifs in proteins. J. Biol. Chem. 273, 31061–31067

Cantrell, D.A. (2001) Phosphoinositide 3-kinase signaling pathways. J. Cell Sci. 114, 1439–1445

Chen, R.Y., Kim, O., Li, M., Xiong, X.S., Guan, J.L., Kung, H.J., Chen, H.G., Shimizu, Y. and Qiu, Y. (2001) Regulation of the PH domain-containing tyrosine kinase Etk by focal adhesion kinase through the FERM domain. Nat. Cell Biol. 3, 439–444

Chung, C.Y., Funamoto, S. and Firtel, R.A. (2001) Signaling pathways controlling cell polarity and chemotaxis. Trends Biochem. Sci. 26, 557–566

Chung, C.Y., Potikyan, G. and Firtel, R.A. (2001) Control of cell polarity and chemotaxis by Akt/PKB and PI3 kinase through the regulation of PAKa. Mol. Cell 7, 937–947

Cockcroft, S. and De Matteis, M.A. (2001) Inositol lipids as spatial regulators of membrane traffic. J. Membr. Biol. 180, 187–194

Comer, F.I. and Parent, C.A. (2002) PI 3-kinases and PTEN: how opposites chemoattract. Cell 109, 541–544

Cozier, G., Lockyer, P.J., Reynolds, J.S., Kupzig, S., Bottomley, J.R., Millard, T., Banting, G. and Cullen, P.J. (2000) GAP1[IP4BP] contains a novel Group I pleckstrin homology domain that directs constitutive plasma membrane association. J. Biol. Chem. 275, 28261–28268

Crompton, A.M., Foley, L.H., Wood, A., Roscoe, W., Stokoe, D., McCormick, F., Symons, M. and Bollag, G. (2000) Regulation of Tiam-1 nucleotide exchange activity by pleckstrin homology domain binding ligands. J. Biol. Chem. 275, 25751–25759

Cullen, P.J. and Venkateswarlu, K. (1999) The potential regulation of ARF6 signalling by phosphatidylinositol 3,4,5-trisphosphate. Biochem. Soc. Trans. 27, 683–690

Cullen, P.J. and Chardin, P. (2000) Membrane targeting—what a difference a G makes. Curr. Biol. 10, R876-R878

Cullen, P.J., Cozier, G.E., Banting, G. and Mellor, H. (2001) Modular phosphoinositide-binding domains—their role in signaling and membrane traffic. Curr. Biol. 11, R882-R893

De Matteis, M.A. and Morrow, J.S. (1998) The role of ankyrin and spectrin in membrane transport and domain formation. Curr. Opin. Cell Biol. 10, 542–549

De Matteis, M.A. and Morrow, J.S. (2000) Spectrin tethers and mesh in the biosynthetic pathway. J. Cell Sci. 113, 2331–2343

De Matteis, M.A., Godi, A. and Corda, D. (2002) Phosphoinositides and the Golgi complex. Curr. Opin. Cell Biol. 14, 434–447

Dhe-Paganon, S., Ottinger, E.A., Nolte, R.T., Eck, M.J. and Shoelson, S.E. (1999) Crystal structure of the pleckstrin homology-phosphotyrosine binding (PH-PTB) targeting region of insulin receptor substrate 1. Proc. Natl. Acad. Sci. U.S.A. 96, 8378–8383

Dowler, S., Currie, R.A., Downes, C.P. and Alessi, D.R. (1999) DAPP1: a dual adaptor for phosphotyrosine and 3-phosphoinositides. Biochem. J. 342, 7–12

Dowler, S., Currie, R.A., Campbell, D.G., Deak, M., Kular, G., Downes, C.P. and Alessi, D.R. (2000) Identification of pleckstrin homology domain-containing proteins with novel phosphoinositide-binding specificities. Biochem. J. 351 19–31

Downing, A.K., Driscoll, P.C., Gout, I., Salim, K., Zvelebil, M.J. and Waterfield, M.D. (1994) Three-dimensional solution structure of the pleckstrin homology domain from dynamin. Curr. Biol. 4, 884–891

Drugan, J.K., Rogers-Graham, K., Gilmer, T., Campbell, S. and Clark, G.J. (2000) The Ras/p120 GTPase-activating protein (GAP) interaction is regulated by the p120 GAP pleckstrin homology domain. J. Biol. Chem. 275, 35021–35027

Farhang-Fallah, J., Yin, XH., Trentin, G., Chengm A.M. and Rozakis-Adcock, M. (2000) Cloning and characterization of PHIP, a novel insulin receptor substrate-1 pleckstrin homology domain interacting protein. J. Biol. Chem. 275, 40492–40497

Federov, A.A., Federov, E., Gertler, F. and Almo, S.C. (1999) Structure of EVH1, a novel proline-rich ligand-binding module involved in cytoskeletal dynamics and neural function. Nat. Struct. Biol. 6, 661–665

Ferguson, K.M., Lemmon, M.A., Schlessinger, J. and Sigler, P.B. (1994) Crystal structure at 2.2 Å resolution of the pleckstrin homology domain from human dynamin. Cell 79 199–209

Ferguson, K.M., Lemmon, M.A., Schlessinger, J. and Sigler, P.B. (1995) Structure of the high affinity complex between inositol 1,4,5-trisphosphate and a phospholipase C pleckstrin homology domain. Cell 83, 1037–1046

Ferguson, K.M., Kavran, J.M., Sankaran, V.G., Fournier, E., Isakoff, S.J., Skolnik, E.Y. and Lemmon, M.A. (2000) Structural basis for discrimination of 3-phosphoinositides by pleckstrin homology domain. Mol. Cell 6, 373–384

Fleming, I.N., Gray, A. and Downes, C.P. (2000) Regulation of the Rac1-specific exchange factor Tiam1 involves both phosphoinositide 3-kinase-dependent and -independent components. Biochem. J. 351, 173–182

Frech, M., Andjelkovic, M., Ingley, E., Reddy, K.K., Falck, J.R. and Hemmings, B.A. (1997) High affinity binding of inositol phosphates and phosphoinositides to the pleckstrin homology domain of RAC/protein kinase B and their influence on kinase activity. J. Biol. Chem. 272, 8474–8481

Franke, T.F., Kaplan, D.R., Cantley, L.C. and Toker, A. (1997) Direct regulation of the Akt proto-oncogene product by phosphatidylinositol 3,4-bisphosphate. Science 275, 665–668

Fukuda, M., Kojima, T., Kabayama, H. and Mikoshiba, K. (1996) Mutation of the pleckstrin homology domain of Bruton's tyrosine kinase in immunodeficiency impaired inositol 1,3,4,5-tetrakisphosphate binding capacity. J. Biol. Chem. 271, 30303–30306

Funamoto, S., Milan, K., Meili, R. and Firtel, R.A. (2001) Role of phosphatidylinositol 3′ kinase and a downstream pleckstrin homology domain-containing protein in controlling chemotaxis in *Dictyostelium*. J. Cell Biol. 153, 795–810

Funamoto, S., Meili, R., Lee, S., Parry, L. and Firtel, R.A. (2002) Spatial and temporal regulation of 3-phosphoinositides by PI 3-kinase and PTEN mediates chemotaxis. Cell 109, 611–623

Fushman, D., Cahill, S., Lemmon, M.A., Schlessinger, J. and Cowburn, D. (1995) Solution structure of pleckstrin homology domain of dynamin by heteronuclear NMR spectroscopy. Proc. Natl. Acad. Sci. U.S.A. 92, 816–820

Fushman, D., Najmabadi-Kaske, T., Cahill, S., Zheng, J., LeVine, H. and Cowburn, D. (1998) The solution structure and dynamics of the pleckstrin homology domain of G protein-coupled receptor kinase 2 (β-adrenergic receptor kinase 1): a binding partner of G$\beta\gamma$ subunits. J. Biol. Chem. 273, 2835–2843

Garcia, P., Gupta, R., Shah, S., Morris, A.J., Rudge, S.A., Scarlata, S., Petrova, V., McLaughlin, S. and Rebecchi, M.J. (1995) The pleckstrin homology domain of phospholipase Cδ1 binds with high affinity to phosphatidylinositol 4,5-bisphosphate in bilayer membranes. Biochemistry 34, 16228–16234

Godi, A., Pertile, P., Meyers, R., Marra, P., DiTullio, G., Iurisci, C., Luini, A., Corda, D. and De Matteis, M.A. (1999) ARF mediates recruitment of phosphatidylinositol 4-monophosphate kinase-β and stimulates synthesis of phosphatidylinositol 4,5-bisphosphate on the Golgi complex. Nat. Cell Biol. 1, 280–287

Hall, A. (1998) Rho GTPases and the actin cytoskeleton. Science 279, 509–514

Hama, H., Schnieders, E.A., Thorner, J., Takemoto, J.Y. and DeWald, D.B. (1999) Direct involvement of phosphatidylinositol 4-phosphate in secretion in the yeast *Saccharomyces cerevisiae*. J. Biol. Chem. 274, 34294–34300

Han, J., Luby-Phelps, K., Das, B., Shu, X., Xia, Y., Mosteller, R.D., Krishna, U.M., Falck, J.R., White, M.A. and Broek, D. (1998) Role of substrates and products of phosphoinositide-3-kinase in regulating activation of Rac-related guanosine triphosphatases by Vav. Science 279, 558–560

Hurley, J.H. and Misra, S. (2000) Signalling and subcellular targeting by membrane binding domains. Annu. Rev. Biophys. Biomol. Struct. 29, 49–79

Hurley, J.H. and Meyer, T. (2001) Subcellular targeting by membrane lipids. Curr. Opin. Cell Biol. 13, 146–152

Hyvonen, M., Macais, M.J., Nilges, M., Oschkinat, H., Saraste, M. and Wilmanns, M. (1995) Structure of the binding site for inositol phosphates in a pleckstrin homology domain. EMBO J. 14, 4676–4685

Hyvonen, M. and Saraste, M. (1997) Structure of the pleckstrin homology domain and Btk motif from Bruton's tyrosine kinase: molecular explanations for X-linked agammaglobulinaemia. EMBO J. 16, 3396–3404

Iijima, M. and Devreotes, P. (2002) Tumor suppressor PTEN mediates sensing of chemoattractant gradients. Cell 109, 599–610

Itoh, T. and Takenawa, T. (2002) Phosphoinositide-binding domains—functional units for temporal and spatial regulation of intracellular signaling. Cell. Signal. 14, 733–743

Jackson, C.L. and Casanova, J.E. (2000) Turning on ARF: the Sec7 family of guanine nucleotide exchange factors. Trends Cell Biol. 10, 60–67

Jiang, Y., Ma, W., Wan, Y., Kozasa, T., Hattori, S. and Huang, X.Y. (1998) The G protein $G\alpha_{12}$ stimulates Bruton's tyrosine kinase and a rasGAP through a conserved PH/BM domain. Nature (London) 395, 808–813

Jin, T., Zhang, N., Long, Y., Parent, C.A. and Devreotes, P.N. (2000) Localization of the G protein $\beta\gamma$ complex in living cells during chemotaxis. Science 287, 1034–1036

Jogl, G., Shen, Y., Gebauer, D., Li, J., Wiegmann, K., Kashkar, H., Kronke, M. and Tong, L. (2002) Crystal structure of the BEACH domain reveals an unusual fold and extensive association with a novel pleckstrin homology domain. EMBO J. 21, 4785–4795

Katanaev, V.L. (2001) Signal transduction in neutrophil chemotaxis. Biochemistry (Mosc) 66, 351–368

Kavran, J.M., Klein, D.E., Lee, A., Falasca, M., Isakoff, S.J., Skolnik, E.Y. and Lemmon, M.A. (1998) Specificity and promiscuity in phosphoinositide binding by pleckstrin homology domains. J. Biol. Chem. 273, 30497–30506

Kimber, W.A., Trinkle-Mulcahy, L., Cheung, P.C.F., Deak, M., Marsden, L.J., Kieloch, A., Watt, S., Javier, R.T., Gray, A., Downes, C.P., Lucocq, J.M. and Alessi, D.R. (2002) Evidence that the tandem-pleckstrin-homology-domain-containing protein TAPP1 interacts with phosphatidylinositol 3,4-bisphosphate and the multi-PDZ-domain-containing protein MUPP1 in vivo. Biochem. J. 361, 525–536

Klarlund, J.K., Guilherme, A., Holik, J.J., Virbasius, A. and Czech, M.P. (1997) Signaling by phosphoinositide 3,4,5-trisphosphate through proteins containing pleckstrin and Sec7 homology domains. Science 275 1927–1930

Klarlund, J.K., Tsiaras, W., Holik, J.J., Chawla, A. and Czech, M.P. (2000) Distinct polyphosphoinositide binding selectivities for pleckstrin homology domains of GRP1-like proteins based on diglycine versus triglycine motifs. J. Biol. Chem. 275, 32816–32821

Klein, D.E., Lee, A., Frank, D.W., Marks, M.S. and Lemmon, M.A. (1998) The pleckstrin homology domain of dynamin isoforms require oligomerization for high affinity phosphoinositide binding. J. Biol. Chem. 273, 27725–27733

Klippel, A., Kavanaugh, W.M., Pot, D. and Williams, L.T. (1997) A specific product of phosphatidylinositol 3-kinase directly activates the protein kinase Akt through its pleckstrin homology domain. Mol. Cell. Biol. 17, 338–344

Koehler, J.A. and Moran, M.F. (2001) RACK1, a protein kinase C scaffolding protein, interacts with the pleckstrin homology domain of p120^{GAP}. Biochem. Biophys. Res. Commun. 283, 888–895

Kojima, T., Fukuda, M., Watanabe, Y., Hamazato, F. and Mikoshiba, K. (1997) Characterisation of the pleckstrin homology domain of Btk as an inositol polyphosphate and phosphoinositide-binding capacity. Biochem. Biophys. Res. Commun. 236, 333–339

Konishi, H., Kuroda, S. and Kikkawa, U. (1994) The pleckstrin homology domain of Rac protein kinase associates with the regulatory domain of protein kinase-C zeta. Biochem. Biophy. Res. Commun. 205, 1770–1775

Konishi, H., Kuroda, S., Tanaka, M., Matsuzaki, H., Ono, Y., Kameyama, K., Haga, T. and Kikkawa, U. (1995) Molecular cloning and characterization of a new member of the RAC protein kinase family: association of the pleckstrin homology domain of three types of RAC protein kinase with protein kinase C subspecies and $\beta\gamma$ subunits of G proteins. Biochem. Biophys. Res. Commun. 216, 526–534

Koshiba, S., Kigawa, T., Kim, J.H., Shirouzu, M., Bowtell, D. and Yokoyama, S. (1997) The solution structure of the pleckstrin homology domain of mouse Son-of-Sevenless 1 (mSos1) J. Mol. Biol. 20, 579–591

Krugmann, S., Anderson, K.E., Ridley, S.H., Risso, N., McGregor, A., Coadwell, J., Davidson, K., Eguinoa, A., Ellson, C.D., Lipp, P., Manifava, M., Ktistakis, N., Painter, G., Thuring, J.W., Cooper, M.A., Lim, Z.Y., Holmes, A.B., Dove, S.K., Michell, R.H., Grewal, A., Nazarian, A., Erdjument-Bromage, H., Tempst, P., Stephens, L.R. and Hawkins, P.T. (2002) Identification of ARAP3, a novel phosphoinositide 3-kinase effector regulating both ARF and Rho GTPases, by selective capture on phosphoinositide affinity matrices. Mol. Cell 9, 95–108

Lemmon, M.A., Ferguson, K.M., O'Brien, R., Sigler, P.B. and Schlessinger, J. (1995) Specific and high-affinity binding of inositol phosphates to an isolated pleckstrin homology domain. Proc. Natl. Acad. Sci. U.S.A. 92, 10472–10476

Lemmon, M.A. and Ferguson, K.M. (2000) Signal-dependent membrane targeting by pleckstrin homology domains. Biochem. J. 350, 1–18

Lemmon, M.A. and Ferguson, K.M. (2001) Molecular determinants in pleckstrin homology domains that allow specific recognition of phosphoinositides. Biochem. Soc. Trans. 29, 377–384

Lemmon, M.A., Ferguson, K.M. and Abrams, C.S. (2002) Pleckstrin homology domains and the cytoskeleton. FEBS Lett. 513, 71–76

Leslie, N.R. and Downes, C.P. (2002) PTEN: The down side of PI 3-kinase signalling. Cell Signal, 14, 285–295

Levine, T.P. and Munro, S. (1998) The pleckstrin homology domain of oxysterol-binding protein recognizes a determinant specific to Golgi membranes. Curr. Biol. 8, 729–739

Levine, T.P. and Munro, S. (2001) Dual targeting of Osh1p, a yeast homologue of oxysterol-binding protein, to both Golgi and the nucleus-vacuole junction. Mol. Biol. Cell 12, 1633–1644

Levine, T.P. and Munro, S. (2002) Targeting of Golgi-specific pleckstrin homology domains involves both phosphatidylinositol 4-kinase-dependent and –independent components. Curr. Biol. 12, 695–704

Li, X., Rivas, M.P., Fang, M., Marchena, J., Mehrotra, B., Chaudhary, A., Feng, L., Prestwich, G.D. and Bankaitis, V.A. (2002) Analysis of oxysterol binding protein homologue Kes1p function in regulation of Sec14p-dependent protein transport from the yeast Golgi complex. J. Cell Biol. 157, 63–78

Lietzke, S.E., Bose, S., Cronin, T., Klarlund, J., Chawla, A., Czech, M.P. and Lambright, D.G. (2000) Structural basis of 3-phosphoinositide recognition by pleckstrin homology domains. Mol. Cell 6, 385–394

Lockyer, P.J., Wennström, S., Venkateswarlu, K., Downward, J. and Cullen, P.J. (1999) Identification of the Ras GTPase-activating protein GAP1^m as an in vivo phosphatidylinositol 3,4,5-trisphosphate-binding protein. Curr. Biol. 9, 265–268

Liu, X., Wang, H., Eberstadt, M., Schnuchel, A., Olejniczak, E.T., Meadows, R.P., Schkeryantz, J.M., Janowick, D.A., Harlan, J.E., Harris, E.A.S., Staunton, D.E. and Fesik, S.W. (1998) NMR structure and mutagenesis of the N-terminal Dbl homology domain of the nucleotide exchange factor Trio. Cell 95, 269–277

Lorra, C and Huttner, W.B. (1999) The mesh hypothesis of Golgi dynamics. Nat. Cell Biol. 1, E113-E155

Lowry, W.E., Huang, J.Y., Lei, M., Rawlings, D. and Huang, X.Y. (2001) Role of the PHTH module in protein substrate recognition by Bruton's agammaglobulinemia tyrosine kinase. J. Biol. Chem. 276, 45276–45281

Ma, A.D., Metjian, A., Bagrodia, S., Taylor, S. and Abrams, C.S. (1998)Cytoskeletal reorganization by G protein-coupled receptors is dependent on phosphoinositide 3-kinase γ, a Rac guanosine exchange factor, and Rac. Mol. Cell. Biol. 18, 4744–4751

Macias, M.J., Musacchio, A., Ponstingi, H., Nigles, M., Saraste, M. and Oschkinat, H. (1994) Structure of the pleckstrin homology domain from β-spectrin. Nature (London) 369, 672–675

Maffucci, T. and Falasca, M. (2001) Specificity in pleckstrin homology (PH) domain membrane targeting: a role for a phosphoinositide-protein co-operative mechanism. FEBS Lett. 506, 173–179

Marshall, A.J., Krahn, A.K., Ma, K.W., Duronio, V. and Hou, S. (2002) TAPP1 and TAPP2 are targets of phosphoinositol 3-kinase signaling in B cells: sustained plasma membrane recruitment triggered by the B cell antigen receptor. Mol. Cell. Biol. 22, 5479–5491

Meili, R., Ellsworth, C., Lee, S., Reddy, T.B., Ma, H. and Firtel, R.A. (1999) Chemoattractant-mediated transient activation and membrane localization of Akt/PKB is required for efficient chemotaxis to cAMP in *Dictyostelium*. EMBO J. 18 2092–2105

Meili, R., Ellsworth, C. and Firtel, R. A. (2000) A novel Akt/PKB-related kinase is essential for morphogenesis in *Dictyostelium*. Curr. Biol. 10, 708–717

Miura, K., Jacques, K.M., Stauffer, S., Kubosaki, A., Zhu, K.J., Hirsch, D.S., Resau, J., Zheng, Y. and Randazzo, P.A. (2002) ARAP1: a point of convergence for ARFs and Rho signaling. Mol. Cell 9, 109–119

Muhlberg, A.B., Warnock, D.E. and Schmid, S.L. (1997) Domain structure and intramolecular regulation of dynamin GTPase. EMBO J. 16, 6676–6683

Nilges, M., Macias, M.J., O'Donoghue, S.I. and Oschkinat, H. (1997) Automated NOESY interpretation with ambiguous distance restraints: The refined NMR solution structure of the pleckstrin homology domain from β-spectrin. J. Mol. Biol. 269, 408–422

Nimnual, A.S., Yatsula, B.A. and Bar-Sagi, D. (1998) Coupling of Ras and Rac guanosine triphosphatases through the Ras exchanger Sos. Science 279, 560–563

Paranavitane, V., Coadwell, J.W., Eguinoa, A., Hawkins, P.T. and Stephens, L.R. (2002) LL5β is a PtdIns(3,4,5)P$_3$-sensor that can bind the cytoskeletal adaptor, γ-filamin. J. Biol. Chem. Papers in Press published October 9 2002 as 10.1074/jbc.M208352200

Parent, C.A., Blacklock, B.J., Froehlich, W.M., Murphy, D.B. and Devreotes, P.N. (1998) G protein signaling events are activated at the leading edge of chemotactic cells. Cell. 95, 81–91

Parent, C.A. and Devreotes, P.N. (1999) A cell's sense of direction. Science 284, 765–770

Peters, P.J., Hsu, V.W., Ooi, C.E., Finazzi, D., Teal, S.B., Ooschot, V., Donaldson, J.G. and Klausner, R.D. (1995) Overexpression of wild-type and mutant ARF1 and ARF6: distinct perturbation of nonoverlapping membrane compartments. J. Cell Biol. 128, 1003–1017

Pitcher, J.A., Touhara, K., Payne, E.S. and Lefkowitz, R.J. (1995) Pleckstrin homology domain-mediated membrane association and activation of the β-adrenergic receptor kinase requires coordinate interaction with G$\beta\gamma$ subunits and lipid. J. Biol. Chem. 270, 11707–11710

Prehoda, K.E., Lee, D.J. and Lim, W.A. (1999) Structure of the enabled/VASP homology 1 domain-peptide complex: a key component in the spatial control of actin assembly. Cell 97, 471–480

Radhakrishna, H., Klausner, R.D. and Donaldson, J.G. (1996) Aluminum fluoride stimulates surface protrusions in cells overexpressing the ARF6 GTPase. J. Cell Biol. 134, 935–947

Radhakrishna, H., Al-Awar, O., Khachikian, Z. and Donaldson, J.G. (1999) ARF6 requirement for Rac ruffling suggests a role for membrane trafficking in cortical actin rearrangements. J. Cell Sci. 112, 855–866

Rameh, L.E. and Cantley, L.C. (1999) The role of phosphoinositide 3-kinase lipid in cell function. J. Biol. Chem. 274, 8347–8350

Raya, A., Revert, F., Navarro, S. and Saus, J. (1999) Characterisation of a novel type of serine/threonine kinase that specifically phosphorylates the human goodpasture antigen. J. Biol. Chem. 274, 12642–12649

Rodriguez, M.M., Ron, D., Touhara, K., Chen, C.-H. and Mochly-Rosen, D. (1999) RACK1, a protein kinase C anchoring protein, coordinates the binding of activated protein kinase C and select pleckstrin homology domains in vitro. Biochemistry 38, 13787–13794

Rossman, K.L., Worthylake, D.K., Snyder, J.T., Siderovski, D.P., Campbell, S.L. and Sondek, J. (2002) A crystallographic view of interactions between Dbs and Cdc43: PH domain-assisted guanine nucleotide exchange. EMBO J. 21, 1315–1326

Russo, C., Gao, Y., Mancini, P., Vanni, C., Porotto, M., Falasca, M., Torrisi, M.R., Zheng, Y. and Eva, A. (2001) Modulation of oncogenic Dbl activity by phosphoinositol phosphate binding to pleckstrin homology domain. J. Biol. Chem. 276 19524–19531

Saito, K., Scharenberg, A.M. and Kinet, J.P. (2001) Interaction between the Bruton's tyrosine kinase pleckstrin homology domain and phosphatidylinositol 3,4,5-trisphosphate directly regulates Btk. J. Biol. Chem. 276, 16201–16206

Sanchez-Madrid, F. and del Pozo, M.A. (1999) Leukocyte polarization in cell migration and immune interactions. EMBO J. 18, 501–511

Santy, L.C. and Casanova, J.E. (2001) Activation of ARF6 by ARNO stimulates epithelial cell migration through downstream activation of both Rac1 and phospholipase D. J. Cell Biol. 154, 599–610

Servant, G., Weiner, O.D., Neptune, E.R., Sedat, J.W. and Bourne, H.R. (1999) Dynamics of a chemoattractant receptor in living neutrophils during chemotaxis. Mol. Biol. Cell 10, 1163–1178

Servant, G., Weiner, O.D., Herzmark, P., Balla, T., Sedat, J.W. and Bourne, H.R. (2000) Polarization of chemoattractant receptor signaling during neutrophil chemotaxis. Science 287, 1037–1040

Shinohara, M., Terada, Y., Iwamatsu, A., Shinohara, A., Mochizuki, N., Higuchi, M., Gotoh, Y., Ihara, S., Nagata, S., Itoh, H., Fukui, Y. and Jessberger, R. (2002) SWAP-70 is a guanine-nucleotide exchange factor that mediates signaling of membrane ruffling. Nature (London) 416, 759–763

Siddhanta, A., Backer, J.M. and Shields, D. (2000) Inhibition of phosphatidic acid synthesis alters the structure of the Golgi apparatus and inhibits secretion in endocrine cells. J. Biol. Chem. 275, 12023–12031

Siddhanta, A., Radulescu, A., Stankewich, M.C., Morrow, J.S. and Shields, D. (2002) Fragmentation of the Golgi apparatus: a role for III spectrin and synthesis of phosphatidylinositol 4,5-bisphosphate. J. Biol. Chem. Published on-line 10.1074/jbc.M209137200

Snyder, J.T., Rossman, K.L., Baumeister, M.A., Pruitt, W.M., Siderovski, D.P., Der, C.J., Lemmon, M.A. and Sondek, J. (2001) Quantitative analysis of the effect of phosphoinositide interactions on the function of Dbl family proteins. J. Biol. Chem. 276, 45868–45875

Snyder, J.T., Worthylake, D.K., Rossman, K.L., Betts, L., Pruitt, W.M., Siderovski, D.P., Der, C.J. and Sondek, J. (2002) Structural basis for the selective activation of Rho GTPases by Dbl exchange factors. Nat. Struct. Biol. 9, 468–475

Soisson, S.M., Nimual, A.S., Uy, M., Bar-Sagi, D. and Kuriyan, J. (1998) Crystal structure of the Dbl and pleckstrin homology domain from the human Son of Sevenless protein. Cell 95, 259–268

Stam, J.C., Sander, E.E., Michiels, F., van Leeuwen, F.N., Kain, H.E.T., van der Kammen, R.A. and Collard, J.G. (1997) Targeting of Tiam1 to the plasma membrane requires the cooperative function of the N-terminal pleckstrin homology domain and an adjacent protein interaction domain. J. Biol. Chem. 272, 28447–28454

Stephens, L.R., Anderson, K.A., Stokoe, D., Erdjument-Bromage, H., Painter, G.F., Holmes, A.B., Gaffney, P.R.J., Reese, C.B., McCormick, F., Tempst, P., Coadwell, J. and Hawkins, P.T. (1998) Protein kinase B kinases that mediate phosphatidylinositol 3,4,5-trisphosphate-dependent activation of protein kinase B. Science 279, 710–714

Stephens, L.R., Ellson, C. and Hawkins, P.T. (2002) Roles of PI3Ks in leukocyte chemotaxis and phagocytosis. Curr. Opin. Cell Biol. 14 203–213

Stites, J., Wessels, D., Uhl, A., Egelhoff, T., Shutt, D. and Soll, D.R. (1998) Phosphorylation of the *Dictyostelium* myosin II heavy chain is necessary for maintaining cellular polarity and suppressing turning during chemotaxis. Cell Motil. Cytoskeleton 39, 31–51

Sweeney, D.A., Siddhanta, A. and Shields, D. (2002) Fragmentation and re-assembly of the Golgi apparatus in vitro. A requirement for phosphatidic acid and phosphatidylinositol 4,5-bisphosphate synthesis. J. Biol. Chem. 277, 3030–3039

Teruel, M.N. and Meyer, T. (2000) Reversible localization of signaling proteins: a dynamic future for signal transduction. Cell 103, 181–184

Thomas, C.C., Dowler, S., Deak, M., Alessi, D.R. and van Aalten, D.M.F. (2001) Crystal structure of the phosphatidylinositol 3,4-bisphosphate-binding pleckstrin homology (PH) domain of tandem PH-domain-containing protein 1 (TAPP1): molecular basis of lipid specificity. Biochem. J. 358, 287–294

Thomas, C.C., Deak, M., Alessi, D.R. and van Aalten, D.M.F. (2002) High-resolution structure of the pleckstrin homology domain of protein kinase B/Akt bound to phosphatidylinositol 3,4,5-trisphosphate. Curr. Biol. 12, 1256–1262

Timm, D., Salim, K., Gout, I., Gurupresad, L., Waterfield, M. and Blundell, T. (1994) Crystal structure of the pleckstrin homology domain from dynamin. Nature (London) Struct. Biol. 1, 782–788

Toker, A. (2002) Phosphoinositides and signal transduction. Cell. Mol. Life Sci. 59, 761–779

Touhara, K., Inglese, J., Pitcher, J.A., Shaw, G. and Lefkowitz, R.J. (1994) Binding of G-protein $\beta\gamma$ subunits to pleckstrin homology domains. J. Biol. Chem. 269, 10217–10220

Tsujishita, Y. and Hurley, J.H. (2000) Structure and lipid transport mechanism of a StAR-related domain. Nat. Struct. Biol. 7, 408–414

Tsukuda, S., Simon, M.I., Witte, O.N. and Katz, A. (1994) Binding of $\beta\gamma$ subunits of heterotrimeric G-proteins to the pleckstrin homology domain of Bruton's tyrosine kinase. Proc. Natl. Acad. Sci. U.S.A. 91, 11256–11260

Vanhaesebroeck, B. and Alessi, D.R. (2000) The phosphoinositide 3-kinase-3-phosphoinositide-dependent kinase 1 connection: more than just a road to protein kinase B. Biochem. J. 346, 561–576

Vanhaesebroeck, B., Leevers, S.J., Ahmadi, K., Timms, J., Katso, R., Driscoll, R.C., Woscholski, R., Parker, P.J. and Waterfield, M.D. (2001) Synthesis and function of 3-phosphorylated inositol lipids. Annu. Rev. Biochem. 70, 535–602

Varnai, P., Rother, K.I. and Balla, T. (1999) Phosphatidylinositol 3-kinase-dependent membrane association of Bruton's tyrosine kinase pleckstrin homology domain visualized in single living cells. J. Biol. Chem. 274, 10983–10989

Venkateswarlu, K., Oatey, P.B., Tavaré, J.M. and Cullen, P.J. (1998a) Insulin-dependent translocation of ARNO to the plasma membrane of adipocytes requires PI 3-kinase. Curr. Biol. 8, 463–466

Venkateswarlu, K., Gunn-Moore, F., Oatey, P.B., Tavaré, J.M. and Cullen, P.J. (1998b) NGF- and EGF-stimulated translocation of the ARF-exchange factor GRP1 to the plasma membrane of PC12 cells requires activation of PI 3-kinase and the GRP1 PH domain. Biochem. J. 335, 139–146

Venkateswarlu, K., Oatey, P.B., Tavaré, J.M., Jackson, T.R. and Cullen, P.J. (1999a) Identification of centaurin-α_1 as an in vivo phosphatidylinositol 3,4,5-trisphosphate-binding protein that is functionally homologous to the yeast ARF GTPase-activating protein Gcs1. Biochem. J. 340, 359–363

Venkateswarlu, K., Gunn-Moore, F., Tavaré, J.M. and Cullen, P.J. (1999b) EGF-stimulated translocation of cytohesin-1 to the plasma membrane of PC12 cells requires PI 3-kinase activation and a functional cytohesin-1 PH domain. J. Cell Sci. 112 1957–1965

Vetter, I.R., Nowak, C., Nishimotot, T., Kuhlmann, J. and Wittinghofer, A. (1999) Structure of the Ran-binding domain complexed with Ran bound to a GTP analogue: implications for nuclear transport. Nature (London) 398, 39–46

Walch-Solimena, C. and Novick, P. (1999) The yeast phosphatidylinositol 4-kinase pik1 regulates secretion at the Golgi. Nat. Cell Biol. 1, 523–525

Waldron, R.T., Iglesias, T. and Rozengurt, E. (1999) The pleckstrin homology domain of protein kinase D interacts preferentially with the eta isoform of protein kinase C. J. Biol. Chem. 274, 9224–9230

Wardlaw, A.J., Brightling, C., Green, R., Woltmann, G. and Pavord, I. (2000) Eosinophils in asthma and other allergic diseases. Br. Med. Bull. 56, 985–1003

Wang, D.S., Shaw, R., Winkelmann, J.C. and Shaw, G. (1994) Binding of pleckstrin homology domains of β-adrenergic receptor kinase and β-spectrin to WD40/β-transducin repeat containing regions of the β-subunit of trimeric G-proteins. Biochem. Biophys. Res. Commun. 203, 29–35

Wang, D.S., Shaw, R., Hattori, M., Arai, H., Inoue, K. and Shaw, G. (1995) Binding of pleckstrin homology domains to WD40/β-transducin repeat containing segments of the protein product of the Lis-1 gene. Biochem. Biophys. Res. Commun. 209, 622–629

Watt, S.A., Kular, G., Fleming, I.N., Downes, C.P. and Lucocq, J.M. (2002) Subcellular localization of phosphatidylinositol 4,5-bisphosphate using the pleckstrin homology domain of phospholipase Cδ1. Biochem. J. 363, 657–666

Watton, S.J. and Downward, J. (1999) Akt/PKB localisation and 3-phosphoinositide generation at sites of epithelial cell-matrix and cell-cell interaction. Curr. Biol. 9, 433–436

Weiner, O.D., Servant, G., Welch, M.D., Mitchison, T.J., Sedat, J.W. and Bourne, H.R. (1999) Spatial control of actin polymerization during neutrophil chemotaxis. Nat. Cell Biol. 1, 75–81

Weiner, O.D. (2002) Regulation of cell polarity during eukaryotic chemotaxis: the chemotactic compass. Curr. Opin. Cell Biol. 14 196–202

Weiner, O.D. (2002) Rac activation: P-Rex1—a convergence point for PtdIns(3,4,5)P_3 and G$\beta\gamma$? Curr. Biol. 12, R429–431

Welch, H.C.E., Coadwell, W.J., Ellson, C.D., Ferguson, G.J., Andrews, S.R., Erdjument-Bromage, H., Tempst, P., Hawkins, P.T. and Stephens, L.R. (2002) P-Rex, a phosphatidylinositol 3,4,5-trisphosphate- and G$\beta\gamma$-regulated guanine nucleotide exchange factor for Rac. Cell 108, 809–821

Wherlock, M. and Mellor, H. (2002) The Rho GTPase family: a Racs to Wrchs story. J. Cell Sci. 115, 239–240

Worthylake, D.K., Rossman, K.L. and Sondek, J. (2000) Crystal structure of Rac1 in complex with the guanine nucleotide exchange region of Tiam1. Nature (London) 408, 682–688

Wyles, J.P., McMaster, C.R. and Ridgway, N.D. (2002) Vesicle-associated membrane protein-associated protein-A (VAP-A) interacts with the oxysterol-binding protein to modify export from the endoplasmic reticulum. J. Biol. Chem. 277, 29908–29918

Xiao, Z., Zhang, N., Murphy, D.B. and Devreotes, P.N. (1997) Dynamic distribution of chemoattractant receptors in living cells during chemotaxis and persistent stimulation. J. Cell Biol. 139, 365–374

Xu, Y.Q., Liu, Y.L., Ridgway, N.D. and McMaster, C.R. (2001) Novel members of the human oxysterol-binding protein family bind phospholipids and regulate vesicle transport. J. Biol. Chem. 276, 18407–18414

Yang, W. and Desiderio, S. (1997) BAP-135, a target for Bruton's tyrosine kinase in response to B cell receptor engagement. Proc. Natl. Acad. Sci. U.S.A. 94, 604–609

Yao, L., Kawakami, Y. and Kawakami, T. (1994) The pleckstrin homology domain of Bruton's tyrosine kinase interacts with protein kinase C. Proc. Natl. Acad. Sci. U.S.A. 91, 9175–9179

Yao, L., Suzuki, H., Ozawa, K., Deng, J., Lehel, C., Fukamachi, H., Anderson, W.B., Kawakami, Y. and Kawakami, T. (1997) Interactions between protein kinase C and pleckstrin homology domains. Inhibition by phosphatidylinositol 4,5-bisphosphate and phorbol 12-myristate 13-acetate. J. Biol. Chem. 272, 13033–13039

Yao, L., Janmey, P., Frigeri, L.G., Han, W., Fujita, J., Kawakami, Y., Apgar, J.R. and Kawakami, T. (1999) Pleckstrin homology domains interact with filamentous actin. J. Biol. Chem. 274 19752–19761

Yoon, H.S., Hajduk, P.J., Petros, A.M., Olejniczak, E.T., Meadows, R.P. and Fesik, S.W. (1994) Solution structure of a pleckstrin homology domain. Nature (London) 369, 672–675

Zhang, P., Talluri, S., Deng, H., Branton, D. and Wagner, G. (1995) Solution structure of the pleckstrin homology domain of $Drosophila\beta$-spectrin. Structure 3, 1185–1195

Zheng, J., Chen, R.-H., Corbalan-Garcia, S., Cahill, S.M., Bar-Sagi, D. and Cowburn, D. (1997) The solution structure of the pleckstrin homology domain of human SOS1: a possible structural role for the sequential association of diffuse B cell lymphoma and pleckstrin homology domains. J. Biol. Chem. 272, 30340–30344

Zhou, M.-M., Ravichandran, K.S., Olejniczak, E.T., Petros, A.M., Meadows, R.P., Sattler, M., Harlan, J.E., Wade, W.S., Burakoff, S.J. and Fesik, S.W. (1995) Structure and ligand recognition of the phosphotyrosine binding domain of Shc. Nature (London) 378, 548–592

Protein Targeting to Endosomes and Phagosomes via FYVE and PX Domains

H. C. G. Birkeland · H. Stenmark

Department of Biochemistry, Norwegian Radium Hospital, Montebello,
0310 Oslo, Norway
E-mail: stenmark@ulrik.uio.no

1	Introduction	90
2	PI3P as a Regulator of Membrane Traffic	91
2.1	Involvement of PI3P in Endocytic, Phagocytic and Autophagic Membrane Traffic	91
2.2	Formation and Turnover of PI3P on Endosomes and Phagosomes	92
3	The PI3P-Binding FYVE Domain	93
3.1	Occurrence of FYVE Domain-Containing Proteins	94
3.2	Structural Basis of FYVE-PI3P Interactions	95
3.3	Functions of FYVE Domain-Containing Proteins on Endosomes and Phagosomes	97
3.3.1	FYVE Domain Proteins in Endocytic Membrane Fusion and Phagosome Maturation	98
3.3.2	FYVE Domain Proteins in Endosomal Protein Sorting	100
3.3.3	FYVE Domain Proteins in Signal Transduction	101
4	The PI3P-Binding PX Domain	101
4.1	Occurrence of PX Domain-Containing Proteins	102
4.2	Structural Basis of PX-PI3P Interactions	103
4.3	Functions of PX Domain-Containing Proteins on Endosomes and Phagosomes	104
4.3.1	PX Domain Proteins in Endosomal Protein Sorting	105
4.3.2	PX Domain Proteins in Signal Transduction	106
4.3.3	PX Domain Proteins in Microbial Killing	106
4.3.4	PX Domain Proteins in Autophagy-Related Processes	107
5	Conclusion	109
References		109

Abstract Phosphatidylinositol 3-phosphate (PI3P) is generated on early endosomal and phagosomal membranes by PI 3-kinases. This lipid serves important regulatory functions in phagocytosis, endocytic traffic, receptor signalling and microbial killing through the recruitment and activation of a number of effector proteins. Almost all of these effectors

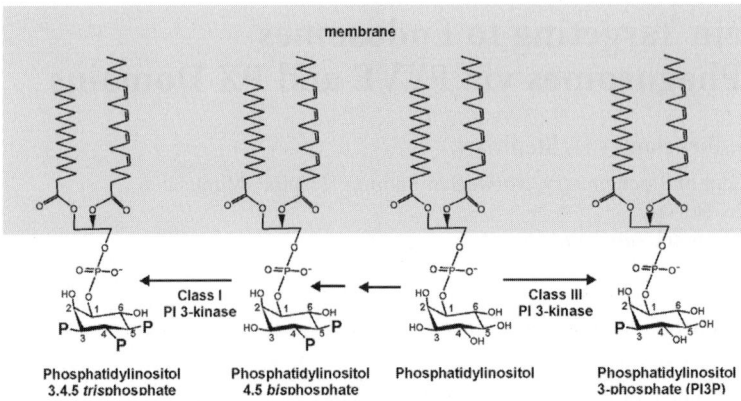

Fig. 1 Formation of PI3P and PI(3,4,5)P$_3$ in cellular membranes. Class I PI 3-kinases are agonist activated, whereas class III PI 3-kinases are independent of ligand stimulation

contain FYVE or PX domains, functional protein modules which are conserved from yeast to mammals. Structural information is available regarding the binding of FYVE and PX domains to PI3P. The two domains are highly different, but they have in common that clusters of basic residues mediate ligand binding through interactions with the phosphate groups of PI3P. Most proteins that contain FYVE or PX domains serve as regulators of endocytic membrane trafficking, whereas others function as regulators of phagosome maturation, signal transduction, microbial killing and other cellular activities of relevance for the immune system.

1
Introduction

Phosphoinositide 3-kinases (PI 3-kinases) regulate a variety of cellular functions through the formation of phosphatidylinositol 3-phosphate (PI3P), phosphatidylinositol 3,4-bisphosphate [PI(3,4)P$_2$] and PI(3,4,5)trisphosphate [PI(3,4,5)P$_3$] (see Fig. 1). Whereas PI(3,4,)P$_2$ and PI(3,4,5)P$_3$ are mainly formed at the plasma membrane (see chapter by Cozier et al., this volume), PI3P is formed on endosomes and phagosomes (Simonsen et al. 2001). There is good evidence that PI3P regulates cellular functions through the specific recruitment of cytosolic proteins that contain FYVE or PX domains to endosomal and phagosomal membranes. In this review, we discuss the mechanisms which account for this recruitment, as well as its functional implications for the ability of the immune system to cope with infectious microorganisms.

2
PI3P as a Regulator of Membrane Traffic

2.1
Involvement of PI3P in Endocytic, Phagocytic and Autophagic Membrane Traffic

Unlike $PI(3,4)P_2$ and $PI(3,4,5)P_3$, the levels of PI3P are rather constant in most cell types, independently of agonist stimulation. The first evidence that PI3P serves as a regulator of membrane traffic was obtained through the studies of yeast *vps* mutants, which display defects in vacuolar protein sorting (the vacuole is the yeast equivalent of a lysosome). One of the Vps proteins, Vps34p, turned out to be a PI3P-forming PI 3-kinase (Schu et al. 1993), and another Vps protein, the protein kinase Vps15p, was found to be a regulatory subunit of this enzyme (Stack and Emr 1994). Homologues of Vps34p and Vps15p are found in mammalian cells (Vanhaesebroeck and Waterfield 1999), and hVPS34 is thought to be the most important (but not exclusive) PI 3-kinase for the formation of PI3P in mammalian cells. Pharmacological inhibitors of PI 3-kinases, such as wortmannin, 3-methyladenine and LY 290004, have been shown to inhibit multiple intracellular trafficking pathways (Simonsen et al. 2001). However, because these inhibitors inhibit not only the class III PI 3-kinase hVPS34, but also class I PI 3-kinases responsible for the formation of $PI(3,4)P_2$ and $(3,4,5)P_3$, the relative contribution of PI3P in membrane trafficking has been difficult to evaluate. Recently, this problem has been overcome through the use of inactivating antibodies to hVPS34 in cell-free assays or by microinjection into intact cells. Such studies have revealed that hVPS34, and hence PI3P, is required for the following trafficking steps in mammalian cells: endocytic membrane fusion, receptor sorting into multivesicular bodies (MVBs), endosome motility, phagosome maturation, and autophagic sequestration (Simonsen et al. 2001). The involvement of PI3P-binding proteins in these trafficking steps is discussed below. We also briefly discuss the possibility that PI3P may regulate other cellular processes, such as growth factor signalling, apoptosis, and assembly of the phagocyte oxidase complex. It should be noted that the disruption of the *Caenorhabditis elegans* homologue of Vps34p causes a complex phenotype which includes an expansion of the nuclear envelope and defective trafficking of a lipoprotein receptor (Roggo et al. 2002). Thus, at least in this organism, PI3P may serve as a regulatory factor of multiple pathways.

2.2
Formation and Turnover of PI3P on Endosomes and Phagosomes

Immunofluorescence and immunoelectron microscopy with a PI3P-specific probe has revealed that PI3P is specifically located to endosomes in fibroblasts (Gillooly et al. 2000). The lipid is found on the limiting membrane of early endosomes and on intraluminal vesicles of MVBs and late endosomes. These vesicles are distinct from those that contain lysobisphosphatidic acid (LBPA), suggesting that PI3P and LBPA follow distinct sorting pathways. In yeast cells, PI3P is likewise found on endosomes and on intraluminal vesicles of the vacuole, the yeast equivalent of late endosomes and lysosomes (Gillooly et al. 2000). Some PI3P is also found on the limiting membrane of the vacuole. PI3P turnover in yeast requires vacuolar hydrolase activity, and in mammalian cells PI3P is more abundant on early than on late endosomes. This indicates that degradation within the lumen of vacuoles and late endosomes/lysosomes represents a major pathway for PI3P turnover.

The above results suggest that PI3P is primarily formed on the limiting membrane of early endosomes. How is this specificity achieved? An important clue came from the observation that p150, the human homologue of Vps15p (the regulatory subunit of the yeast PI 3-kinase Vps34p), interacts with the active, GTP-bound form of the small GTPase Rab5, which is localized to early endosomes (Christoforidis et al. 1999b). This observation suggested the possibility that the hVPS34/p150 complex might be specifically targeted to early endosomes by Rab5, and recent fluorescence microscopy studies support this idea (Murray et al. 2002).

PI3P formed on endosome membranes appears to be turned over at a fast rate (Pattni et al. 2001). Genetic studies in yeast (Wurmser and Emr 1998) and immunoelectron microscopic studies of mammalian cells (Gillooly et al. 2000) indicate that internalisation into multivesicular bodies (MVBs) represents a major pathway of PI3P turnover. Presumably, when MVBs fuse with lysosomes/vacuoles, their intraluminal vesicles are degraded by lipases (Stenmark and Gillooly 2001). However, specific phosphatases for PI3P are conserved from yeast to human and also play a role in PI3P turnover (Taylor et al. 2000). Interestingly, mutations in some of the corresponding genes are associated with human disease (Wishart and Dixon 2002; Laporte et al. 2002).

Although PI3P production on endosomes appears to be constitutive, PI3P formation on phagosomal membranes is strongly induced on the engulfment of particles (Ellson et al. 2001a). As soon as the early phagosome has been pinched off from the plasma membrane, a high production of PI3P can be detected by live-cell fluorescence microscopy with PI3P-specific probes (Ellson et al. 2001a; Pattni et al. 2001). Because $PI(3,4,5)P_3$ is formed at the phagosomal cup (Vieira et al. 2001), PI3P might, in principle, be formed by dephosphorylation of this phosphoinositide. However, microinjected antibodies against hVPS34 effectively block the accumulation of PI3P on early phagosomes, indicating that phagosomal PI3P is mainly formed by phosphorylation of phosphatidylinositol by hVPS34 (Vieira et al. 2001). It is not known how hVPS34 is targeted to early phagosomes, but as these structures contain Rab5 (Garin et al. 2001) it is reasonable to speculate that the mechanism is similar to the targeting of hVPS34 to early endosomes. Once formed at the early phagosome membrane, PI3P is present for several minutes until it eventually disappears, possibly through the activity of PI 3-phosphatases (Vieira et al. 2001).

The finding that PI3P is an important regulator of membrane trafficking raised the possibility that it might recruit specific cytosolic proteins. A number of such PI3P effectors have now been identified, almost all of which are distinguished by the presence of FYVE or PX domains.

3
The PI3P-Binding FYVE Domain

The FYVE domain (conserved in Fab1p, YOTB, Vac1p and EEA1) was originally identified as a zinc finger required for the localization of the early endosomal autoantigen EEA1 to endosomal membranes (Stenmark et al. 1996). Subsequent studies revealed that the FYVE domain of EEA1, as well as those of a number of other proteins, bind with high specificity to PI3P (Stenmark et al. 2002; Sankaran et al. 2001).

Table 1 Examples of mammalian FYVE domain-containing proteins (for a more exhaustive list, see Stenmark et al. 2002)

Protein	Proposed function	Localisation
EEA1	Rab5 effector. Tethering of endosomal membranes. Phagosome maturation	Early endosomes. Early phagosomes
Rabenosyn-5	Rab4/Rab5 effector. Tethering of endosomal membranes. Organisation of endosomal microdomains.	Early endosomes
Rabip4	Rab4 effector. Endocytic recycling	Early endosomes
Hrs	Sorting of ubiquitinated membrane proteins from endosomes to lysosomes.	Early endosomes
SARA	Scaffolding of Smad2 in TGF-β signalling	Early endosomes
MTMR3	PI3P and Pl(3,5)P$_2$ 3-phosphatase. Regulator of autophagy	Cytosol, autophagosomes?
PIKfyve	PI and PI3P 5-kinase. MVB formation	Early endosomes
Fgd1[a]	GDP/GTP exchange factor for Cdc42. Regulation of the subcortical actin cytoskeleton	Plasma membrane

[a] The FYVE domain of Fgd1 has low affinity for PI3P and also binds PI5P (Sankaran et al. 2001). See text for references.

3.1
Occurrence of FYVE Domain-Containing Proteins

The major hallmarks of the FYVE domain are eight conserved cysteines and the sequence R(R/K)HHCRXCG (in single-letter amino acid code; 'X' indicates any amino acid) that surrounds the third cysteine. While the conserved cysteines co-ordinate two Zn^{2+} ions that stabilise the FYVE structure, basic residues of the R(R/K)HHCRXCG motif are directly involved in ligand binding (see below). FYVE domains are not found in prokaryotes, but they are conserved in eukaryotes from yeast to human. The frequency of FYVE domain-encoding genes in various organisms is about 1 FYVE domain per 1,000 genes. With a couple of exceptions, FYVE domains occur in one copy per protein. From the human genome sequence, 27 different FYVE domain containing proteins can be predicted. Most of these have been cloned and characterised to some extent (some examples are given in Table 1). In all cases tested, FYVE domains bind PI3P, although the affinity varies (Gillooly et al. 2001a; Gaullier et al. 1998).

In general, FYVE domain-containing proteins have little in common, except that they are cytosolic proteins. Different FYVE proteins may contain a variety of different structural elements in addition to the FYVE domain, including coiled-coil regions, RUN domains, ankyrin repeats, WD40 repeats and PH domains (Stenmark et al. 2002). The subgroup of FYVE proteins that contain PH domains have a weak affinity for PI3P, and they mainly localise to the plasma membrane. For these proteins, it is conceivable that their membrane recruitment is mediated by their PH domains and not by their FYVE domain. In most other cases tested, FYVE proteins localise to early endosomes. A few FYVE domain proteins have catalytic domains, including phosphatase, kinase and GDP/GTP exchange factor domains. It is interesting to note that one FYVE domain protein, Fab1p/PIKfyve, is a PI3P 5-kinase (Odorizzi et al. 1998; Sbrissa et al. 1999) whereas another one, MTMR3, is a PI3P phosphatase (Taylor et al. 2000; Walker et al. 2001). Perhaps the FYVE domains of these proteins serve to recruit them to membranes that contain their substrate, PI3P.

3.2
Structural Basis of FYVE-PI3P Interactions

The FYVE domain is a compact ~60-residue structure that is stabilised by two Zn^{2+} ions (Stenmark et al. 1996; Misra and Hurley 1999). Three FYVE domain structures have been solved by X-ray crystallography so far; those of yeast Vps27p, *Drosophila* Hrs and human EEA1 (Misra and Hurley 1999; Mao et al. 2000; Dumas et al. 2001). These three structures have a highly similar fold consisting of a pair of two-stranded β sheets and a C-terminal α helix. Eight conserved cysteines coordinate the two Zn^{2+} ions in a cross-braced topology (in Vps27p, the sixth cysteine is replaced by a histidine). The general fold of the FYVE domain is very similar to those of two other zinc fingers, the PHD and RING fingers, but the conserved R(R/K)HHCRXCG motif is unique to the FYVE domain. According to the EEA1 FYVE crystal structure, which is the only one that has been obtained in the ligand bound form (Dumas et al. 2001), all the basic residues of this motif are involved in ligand binding. Specifically, the arginine-arginine (or -lysine) pair mediate co-ordination of the 1-phosphate of PI3P, whereas the two histidines and the third arginine co-ordinate the 3-phosphate group. In addition, the second histidine forms contact with inositol hydroxyl groups, and a C-terminally located arginine [not part of the R(R/K)HHCRXCG motif] also contributes to the co-ordination of the 3-phosphate group. Side chains that interact with inositol hydroxyl groups block spaces which would have been occupied by 4- or 5-phosphates. This could explain the exquisite specificity of the FYVE domain for PI3P.

The most striking difference between the three published crystal structures of FYVE domains is that Vps27p FYVE was solved as a monomer (Misra and Hurley 1999), whereas both Hrs FYVE and EEA1 FYVE were solved as dimers. The structure of Hrs FYVE showed a dimer which was held together by citrate ions from the crystallization medium (Mao et al. 2000). Under the assumption that citrate may mimic PI3P, it was proposed that the interface between the two FYVE monomers creates two ligand binding sites. However, the crystal structure of EEA1 FYVE complexed to PI3P indicates that this model for ligand binding is not correct (Dumas et al. 2001). Even though EEA1 FYVE was also solved in dimeric form, this dimer configuration differed radically from that of Hrs FYVE. With EEA1 FYVE, dimerisation was mainly caused by parallel contacts between N-terminal coiled-coil domains, and each monomer was found to contain an independent ligand binding site. The coiled-coil dimerisation and the binding to the ligand strongly restrict the orientation of the EEA1 FYVE dimer with respect to the membrane, so a model for the membrane association of EEA1 FYVE can be predicted with some confidence (Dumas et al. 2001) (see Fig. 2). According to this model, membrane association is mediated by the interaction of basic patches from the two FYVE domains with individual PI3P headgroups and strengthened by the insertion of a hydrophobic loop into the membrane interface. In agreement with this model, FYVE domains bind with much higher affinity to membrane-associated PI3P than to the free headgroup (Gaullier et al. 2000; Sankaran et al. 2001). Some but not all FYVE domain proteins contain predicted coil-regions, but it is not clear whether dimerisation is a universal requirement for the binding of FYVE proteins to membranes. Although a monomeric FYVE domain from Hrs does not bind to endosome membranes, a construct consisting of two Hrs FYVE domains in tandem is strongly targeted to endosomes (Gillooly et al. 2000). This suggests that FYVE dimers may bind strongly to PI3P-containing membranes via an avidity effect. EEA1 and several other FYVE proteins interact with the GTP-bound form of endosomal GTPases, and the membrane association of these proteins is likely to be mediated by a dual interaction with PI3P and the Rab GTPase (Simonsen et al. 1998).

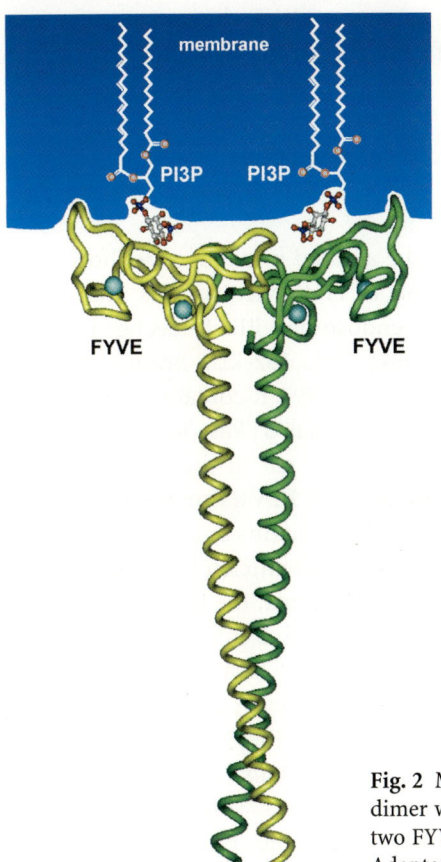

Fig. 2 Model for the interaction of an EEA1 dimer with a PI3P-containing membrane, via two FYVE domains (Dumas et al. 2001). Adapted from (Stenmark et al. 2002). Zn^{2+} ions are indicated in *light blue*

3.3
Functions of FYVE Domain-Containing Proteins on Endosomes and Phagosomes

Most of the FYVE proteins that have been studied to date are localised to early endosomes (see Table 1). This is not very surprising, given that PI3P is enriched on the limiting membrane of these organelles. The early endosome is an important sorting station for endocytic trafficking and is also thought to play a role in compartmentalised receptor signalling. Accordingly, several FYVE domain proteins function as regulators of endocytic membrane traffic, and at least one FYVE domain protein is a mediator of transforming growth factor-β (TGF-β) signalling.

3.3.1
FYVE Domain Proteins in Endocytic Membrane Fusion and Phagosome Maturation

The first group of FYVE proteins to be studied in detail were those concerned with the tethering and fusion of endocytic vesicles and endosomes. A well-known marker of early endosomes, EEA1, has been particularly well characterised in this context. This 162-kDa protein forms a parallel coiled-coil dimer (Callaghan et al. 1999). It contains two binding sites for the GTP-bound form of the endosomal GTPase Rab5, one at the N-terminus, and one at the C-terminus adjacent to the FYVE domain (Simonsen et al. 1998). Although the isolated FYVE domain of EEA1 is not targeted to endosomes, a construct that contains both the C-terminal Rab5 binding domain and the FYVE domain is efficiently recruited onto early endosomes (Stenmark et al. 1996). Moreover, the targeting of EEA1 to endosomes is inhibited by dominant-negative Rab5 (Raiborg et al. 2001b). This argues that endosomal targeting of EEA1 is mediated by a dual interaction with Rab5:GTP and PI3P. However, recent results indicate that, whereas the isolated FYVE domain of EEA1 is monomeric, the construct which contains the Rab5 binding domain as well forms a coiled-coil dimer (Dumas et al. 2001) (see Fig. 2). Therefore, the ability of this construct to be targeted to early endosomes may alternatively be caused by its potential binding to two PI3P molecules, thus increasing its avidity for PI3P-containing membranes. The requirement for Rab5 might therefore be explained by its role in PI3P formation (Christoforidis et al. 1999b). Although the binding to PI3P could mediate the principal recruitment of EEA1 to early endosomes, Rab5 binding might be required to recruit EEA1 to specific microdomains that contain the machinery for membrane docking and fusion.

Immunodepletion of EEA1 from cytosol inhibits homotypic fusion between early endosomes in vitro, and it also inhibits the heterotypic fusion between endocytic vesicles and early endosomes (Simonsen et al. 1998; Rubino et al. 2000). This indicates that EEA1 is an essential component of the machinery that mediates homo- and heterotypic endocytic membrane fusion controlled by Rab5. There is evidence that EEA1 is required before SNARE-mediated docking and fusion, suggesting that

EEA1 functions as a tethering molecule (Christoforidis et al. 1999a). Because EEA1 is a long, rod-shaped coiled-coil dimer with Rab5 binding sites at both the N- and C-termini, it would seem suited for tethering two Rab5-positive membranes, such as an endocytic vesicle and an early endosome. Chemical cross-linking experiments suggest that membrane-associated EEA1 forms multimeric complexes which contain the endosomal SNARE protein syntaxin13 (McBride et al. 1999), and EEA1 has also been shown to interact with another endosomal SNARE protein, syntaxin6 (Simonsen et al. 1999). It is conceivable that the multimerisation and interactions with SNARE proteins are important for the ability of EEA1 to mediate membrane fusion, although the molecular details are still not understood.

Even though EEA1 has been mainly studied in the context of endocytic membrane fusion, its function is not confined to that of endocytosis. EEA1 is abundant on early phagosomes (Garin et al. 2001; Fratti et al. 2001a), and recent evidence suggests that it serves to regulate phagosome maturation. Microinjection of anti-EEA1 antibodies inhibits the ability of latex bead phagosomes to acquire late phagosomal markers (Fratti et al. 2001b), suggesting that EEA1 is essential for this process. An EEA1-interacting SNARE protein, syntaxin6, is also essential for phagosome maturation. Syntaxin6 is found in the TGN and on early endosomes/phagosomes, and it is possible that this SNARE protein mediates TGN to endosome/phagosome trafficking. As such trafficking is required for phagosome maturation, EEA1 might be required for tethering syntaxin6-containing TGN-derived vesicles to early phagosomes (Gillooly et al. 2001b). In addition to binding Rab5, the N-terminal C2H2 zinc finger of EEA1 also binds Rab22, a Rab GTPase found both in the Golgi and on endosomes (Kauppi et al. 2002). It will be interesting to study the possible role of Rab22 in phagosome maturation.

Several other FYVE domain-containing proteins have been implicated in endocytic membrane fusion, and like EEA1 these proteins are Rab effectors. Rabenosyn-5 contains two distinct Rab5 binding sites, like EEA1, and in addition this protein contains a Rab4 binding domain (Nielsen et al. 2000; de Renzis et al. 2002; Merithew et al. 2002). Like EEA1, Rabenosyn-5 is required for endocytic membrane fusion. Moreover, overexpression of this protein causes an increased association between Rab4- and Rab5-positive endosomal microdomains (de Renzis et al. 2002). These results suggest that Rabenosyn-5 may function both as a tethering molecule and as an organizer of membrane microdomains.

Consistent with the former function, there is good evidence that the putative yeast homologue of Rabenosyn-5, Vac1p, plays a role in vesicle docking (Peterson et al. 1999; Tall et al. 1999). Another protein that causes the merging of Rab4-and Rab5-containing microdomains has also been identified. This protein, Rabip4, has been shown to interact with Rab4 (Mari et al. 1999). Overexpression of Rabip4 causes an intracellular retention of the glucose transporter Glut1, suggesting that it may function as a regulator of endocytic recycling.

3.3.2
FYVE Domain Proteins in Endosomal Protein Sorting

From early endosomes, transmembrane proteins may be sorted to several destinations, including the plasma membrane and lysosomes. Sorting to lysosomes, which occurs through the formation of MVBs, requires specific signals, and mono-ubiquitin functions as such a signal when conjugated to the cargo (Hicke 2001). Recent evidence indicates that a FYVE protein, Hrs (hepatocyte growth factor regulated tyrosine kinase substrate), is an important component of the machinery which sorts ubiquitinated membrane proteins from endosomes to lysosomes. Hrs and its yeast homologue, Vps27p, sort ubiquitinated cargo via their ubiquitin-interacting motifs (UIMs) (Raiborg et al. 2002; Bilodeau et al. 2002; Shih et al. 2002). The C-terminus of Hrs contains a clathrin binding domain (Raiborg et al. 2001a), and Hrs is found in a characteristic bilayered clathrin coat on early endosomes (Raiborg et al. 2002; Sachse et al. 2002). Whereas overexpression of Hrs causes an increased recruitment of clathrin to endosomes (Raiborg et al. 2001a), wortmannin treatment, which causes Hrs to dissociate from endosome membranes (Urbé et al. 2000; Raiborg et al. 2001b), depletes clathrin from endosomes (Raiborg et al. 2001a; Sachse et al. 2002). This suggests that Hrs may function as a clathrin adapter for ubiquitinated membrane proteins. The function of the Hrs-containing clathrin coat is not known, but it might serve to concentrate cargo in specific endosomal microdomains for further trafficking into MVBs (Raiborg and Stenmark 2002).

3.3.3
FYVE Domain Proteins in Signal Transduction

Accumulating evidence suggests that growth factor receptor signalling not only occurs from the plasma membrane but also from internal membranes such as endosomes (Ceresa and Schmid 2000; Sorkin and Von Zastrow 2002). Signalling complexes that assemble on endosome membranes might be distinct from those formed at the plasma membrane, which could be important for fine tuning growth factor responses. TGF-β, a vital regulator of cell growth and differentiation, appears to mediate signalling through Smad2 only after endocytosis of the ligand/receptor complex (Hayes et al. 2002). Interestingly, a FYVE protein, SARA (Smad anchor for receptor activation), is important for TGF-β signalling through Smad2. SARA localises to early endosomes via its FYVE domain (Panopoulou et al. 2002; Hayes et al. 2002; Itoh et al. 2002) and probably functions as a scaffold which presents Smad2 to the ligand-bound receptor complex (Tsukazaki et al. 1998). The receptor kinase domain then phosphorylates Smad2, which then associates with Smad4 to form a complex that enters the nucleus and functions as a transcription factor.

4
The PI3P-Binding PX Domain

The PX (Phox homology) domain was identified as an orphan domain present in several proteins of biological interest, including two subunits of the phagocyte oxidase (Phox) complex (Ponting 1996). The recent discovery that the PX domain binds phosphoinositides (Kanai et al. 2001; Song et al. 2001; Cheever et al. 2001; Ellson et al. 2001b) has provided clues about possible localisations and functions of a number of proteins. Most PX domains studied to date, including all those found in yeast (Yu and Lemmon 2001), bind specifically to PI3P, but there are also a few examples of PX domains with different specificities (Sato et al. 2001; Ellson et al. 2002). Here we focus on the PI3P specific PX domains [plus one example of a PI(3,4)P_2 binding PX domain].

Table 2 Examples of mammalian PX domain-containing proteins (for a more exhaustive list, see Ellson et al. 2002; see text for references)

Protein	Proposed function	Localisation
Snx1	Sorting of EGF receptors from early endosomes to lysosomes.	Early endosomes.
Snx3	Sorting of receptors from early to recycling endosomes	Early endosomes
CISK	Ser/Thr protein kinase. Anti-apoptotic signalling downstream of IGF-I and EGF receptors	Early endosomes
p40^{phox}	Subunit of phagocyte oxidase complex	Cytosol. Early endosomes. Early phagosomes?
p47^{phox}	Subunit of phagocyte oxidase complex	Cytosol. Early endosomes. Phagosomal cup?

4.1
Occurrence of PX Domain-Containing Proteins

The PX domain is about 120 residues long and contains several characteristic sequences, including a number of basic residues and a proline-rich stretch. The sequence identities between distinct PX domains are lower than those between distinct FYVE domains. Like FYVE domains, PX domains are absent from prokaryotic proteins but present in eukaryotes from yeast to human (some examples of mammalian PX domain-containing proteins are provided in Table 2). The PX domain is roughly twice as abundant as the FYVE domain in various organisms. Like the FYVE domain, it usually occurs in one copy per protein. It may co-occur with a number of other domains, such as SH3 domains, C2 domains, RUN domains, FERM domains and spectrin repeats (Ellson et al. 2002). Some PX domain proteins contain catalytic domains, for instance, protein kinase regions and kinesin motor domains. A few PX domain proteins contain transmembrane domains, whereas most PX domain proteins shuttle between cytosol and endosome membranes.

4.2
Structural Basis of PX-PI3P Interactions

The structures of the PI3P-binding PX domains Vam7p and p40[phox] have been solved by NMR spectroscopy and X-ray diffraction spectroscopy, respectively (Cheever et al. 2001; Bravo et al. 2001), whereas both the solution and crystal structures of the PI(3,4)P$_2$-binding PX domain of p47[phox] have been solved (Hiroaki et al. 2001; Karathanassis et al. 2002). These analyses show that the PX domain structure is wedge shaped and consists of a three-stranded β-sheet followed by three α-helices. The high-resolution crystal structure of the ligand-bound PX domain of p40[phox] shows that it is the α-helical face of the wedge which contains the ligand-binding pocket. A loop between the two first α-helices is predicted to insert into the membrane, and basic residues located within α1, the membrane-inserting loop and α2 contact the phosphates of PI3P. A tyrosine residue located in α1 stacks against the inositol ring of PI3P and thus forms the floor of the ligand-binding pocket. It is interesting to note that a conserved polyproline motif (PxxP) is present in the membrane interaction loop of many PX domains, including p40[phox] and p47[phox]. This suggests the possibility that the membrane interactions of PX domains might be regulated by proteins which contain polyproline-binding domains, such as SH3 domains. Indeed, NMR analysis of p40[phox] indicates that the PX domain of this protein binds to its own C-terminal PxxP motif (Hiroaki et al. 2001). However, the crystal structure of the ligand-bound PX domain of p40[phox] indicates that the PxxP motif is buried when ligand is bound (Bravo et al. 2001). Thus if the PXXP motif of p40[phox] does interact *in cis* with the SH3 domain, this can only be true in the absence of ligand and one would have to postulate large conformational differences between the unbound and ligand-bound form. Interestingly, p47[phox] binds to PI(3,4)P$_2$-containing membranes only when the intramolecular association between its PX domain and its C-terminal SH3 domain is abolished by phosphorylation (Karathanassis et al. 2002) (see Fig. 3). The ability of PX domains to undergo regulated conformational changes and membrane binding is likely to be important for their functions (see Sect. 4.3.3).

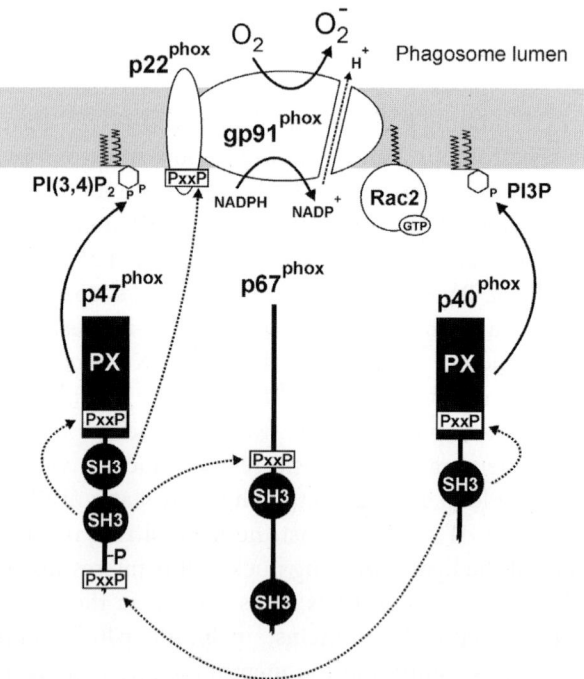

Fig. 3 Model for the activation of the Phox complex through recruitment of cytosolic components, including the two PX domain proteins p47^{phox} and p40^{phox}. Translocation of the cytosolic Phox subunits to phagosome membranes probably involves phosphoinositide recognition by PX domains (*solid arrows*) as well as the rearrangements of protein–protein interactions mediated by SH3 domains and PxxP motifs. Examples of the latter interactions are indicated by *dashed arrows*. Phosphorylation of p47^{phox} prevents the interactions between its own SH3 domain and PxxP motif. This might enable the PX domain to interact with PI(3,4)P$_2$ (formed in the phagosomal cup) and the SH3 domain to interact with a PxxP sequence in p22^{phox}. p67^{phox} and p40^{phox} might be recruited through their interactions with p47^{phox} and the interaction of p47^{phox} with PI3P in the phagosome membrane. (Adapted from Wishart et al. 2001)

4.3
Functions of PX Domain-Containing Proteins on Endosomes and Phagosomes

Like FYVE domain proteins, many PX domain proteins regulate endosomal membrane trafficking (Ellson et al. 2002; Sato et al. 2001). However, their exact functions in membrane trafficking are distinct from those of

FYVE domain proteins. One yeast PX domain protein, Vam7p, is a SNARE molecule involved in membrane fusion with the vacuole (there are no direct mammalian homologues of Vam7p) (Cheever et al. 2001). A number of other PX domain proteins appear to function in protein sorting out of early endosomes, albeit in a different manner than the FYVE domain protein Hrs. There is also one well-characterised example of a PX domain protein which functions in signal transduction, and this protein has a completely different function than the signal-transducing FYVE domain protein SARA. Moreover, in yeast, two PX domain proteins are required for the cytoplasm to vacuole targeting (Cvt) pathway, an autophagy-related biosynthetic process.

4.3.1
PX Domain Proteins in Endosomal Protein Sorting

The sorting nexin (Snx) family is the largest family of PX domain proteins, with 25 members in humans (Teasdale et al. 2001; Worby and Dixon 2002). Their PX domains form a subgroup among PX domains, and in general they localise to early endosomes (Teasdale et al. 2001). Snx1 was identified in a screen for proteins which interact with a portion of the cytoplasmic tail of the epidermal growth factor receptor that is essential for its endosomal sorting to lysosomes (Kurten et al. 1996). Because overexpression of Snx1 causes mis-sorting of the epidermal growth factor receptor, it was concluded that Snx1 is involved in receptor sorting, hence the term "sorting". Subsequent work revealed that several other members of the sorting nexin family also bind to and cause mis-sorting of receptors when overexpressed, and Snx1, Snx3 and Snx15 affect endosome morphology at high expression levels (Cozier et al. 2002; Xu et al. 2001; Barr et al. 2000). Microinjected antibodies against Snx3 inhibit the transport of internalised transferrin from sorting to recycling endosomes (Xu et al. 2001), suggesting that this protein may regulate transferrin receptor recycling. However, so far, the exact functions of Snx proteins are not known. It is interesting to note that Snx1 has been found to interact both with Snx2 and with the FYVE domain protein Hrs (Haft et al. 2000; Chin et al. 2001), although the functional implications of these interactions remain to be sorted out. In any case, there is no doubt that Snx proteins are physiologically very important: Whereas $Snx1^{-/-}$ mice and $Snx2^{-/-}$ mice are healthy and viable, the double knockout is embryonic lethal (Schwarz et al. 2002).

The strongest evidence for a role of Snx family members in protein sorting comes from the yeast *S. cerevisiae*. The two Snx family proteins Vps5p (putative homologue of Snx1) and Vps17p are essential for normal vacuolar protein sorting. These proteins are subunits of a large protein complex called the "retromer", which associates with cytoplasmic tails of membrane proteins and mediates their recycling from endosomes to the Golgi complex (Seaman et al. 1998; Sato et al. 2001). There is evidence that retromer function requires the formation of PI3P (Burda et al. 2002), but further studies are needed to clarify the exact biochemical functions of the retromer and other Snx protein-containing complexes.

4.3.2
PX Domain Proteins in Signal Transduction

Several cell types undergo apoptosis in the absence of serum. The cytokine-independent survival kinase CISK was identified as a protein which counteracts apoptosis induced by withdrawal of interleukin-3 (Liu et al. 2000). The kinase domain of CISK resembles that of another anti-apoptotic kinase, protein kinase B (PKB), whose involvement in signalling cascades downstream of the insulin receptor has been thoroughly characterised (Chan et al. 1999). However, whereas PKB is recruited to the plasma membrane, CISK can be found on early endosomes (Virbasius et al. 2001). The differential targeting of PKB and CISK can be explained by their different N-termini. PKB contains an N-terminal PH domain which binds to $PI(3,4)P_2$ and $PI(3,4,5)P_3$, formed at the plasma membrane on agonist stimulation. In contrast, CISK contains an N-terminal PX domain which has been found to bind PI3P (Virbasius et al. 2001). CISK is likely to phosphorylate many of the same apoptotic regulators as PKB, although this still remains to be studied. As CISK is activated through IGF-1 and EGF receptors, these receptors may have a mode of anti-apoptotic signalling which occurs from endosomes.

4.3.3
PX Domain Proteins in Microbial Killing

The Phox (or NADPH oxidase) complex found in neutrophilic granulocytes generates superoxide anions which are used to kill phagocytosed microorganisms (Babior 1999; Vignais 2002). Dysfunctional Phox complexes give rise to chronic granulomatous diseases, genetic disorders characterised by severe and recurrent infections. The complex consists of an integral membrane component, flavocytochrome b_{558} (which consists of the subunits gp91^{phox} and p22^{phox}) and four cytosolic subunits, Rac2 (a small GTPase), p67^{phox}, p47^{phox} and p40^{phox}. On stimulation of the phagocyte, the cytosolic subunits translocate to membranes and the Phox complex is activated. Such stimulation involves a number of signals, including PI 3-kinase activation (Babior 1999; Ellson et al. 2001b). Both p47^{phox} and p40^{phox} contain PX domains which might target these subunits to the appropriate membranes. The PX domain of p47^{phox} is atypical in the sense that it binds preferentially to PI(3,4)P$_2$, a phosphoinositide which is generated in the phagocytic cup (Marshall et al. 2001). On the other hand, the PX domain of p40^{phox} binds strongly to PI3P and targets this subunit to early endosomes when expressed in non-phagocytic cells (Ellson et al. 2001b). Because early phagosomes contain PI3P (Vieira et al. 2001), p40^{phox} is also likely to be recruited to these organelles. This argues that p47^{phox} and p40^{phox} may be recruited sequentially onto the forming phagosome in order to assemble a functional Phox complex. In addition to recruiting p47^{phox} and p40^{phox} to membranes, phosphoinositides may also serve to sterically activate these proteins, and their activation also involves phosphorylation. As discussed in Sect. 4.2, phosphorylation of p47^{phox} opens up its conformation by releasing the *cis*-interaction between its PX and SH3 domains. This could enable the PX domain to interact with PI(3,4)$_2$ and the SH3 domain to interact with a PXXP sequence within p22^{phox} (see Fig. 2) (Ellson et al. 2002; Wishart et al. 2001).

4.3.4
PX Domain Proteins in Autophagy-Related Processes

The cytoplasm to vacuole targeting (Cvt) pathway in yeast is related to starvation-induced autophagy but functions constitutively under vegetative growth conditions. This constitutive pathway is used to deliver specific cytosolic components, such as aminopeptidase I (ApeI), to the vacuole lumen (Klionsky et al. 1992). The Cvt pathway is mechanistically similar to the autophagy pathway as it leads to sequestration of cytoplasmic cargo through the formation of a double membrane transport intermediate, the Cvt vesicle, which then fuses with the vacuole (Scott and Klionsky 1998). The majority of the proteins required for the Cvt and autophagy pathways localise to the same perivacuolar pre-autophagosomal structure (PAS), and even though there is significant overlap between the proteins involved in the two pathways, pathway-specific proteins also exist. For instance, the Cvt-specific Cvt19p receptor protein has been shown to select cargo which utilises the Cvt pathway (Scott et al. 2001; Leber et al. 2001), leading to cargo incorporation into the Cvt vesicle through Cvt19p's interaction with Apg8p, an autophagy protein which localises to the PAS, where Cvt vesicle formation may take place (Shintani et al. 2002).

Recently, it was shown that Vps34p PI 3-kinase is required for constitutive autophagy by the Cvt pathway as well as in starvation-induced autophagy in yeast (Wurmser and Emr 2002). Three PI3P-binding proteins have been suggested to be PI3-kinase effectors in the Cvt pathway. The first two proteins are the sorting nexin-like proteins Cvt13p and Cvt20p, which have both been found to be required for transport of ApeI through the Cvt pathway (Nice et al. 2002). Both proteins contain PX domains which bind PI3P, and their PX domains are necessary for their localisation to the PAS and for Cvt pathway function. Cvt13p and Cvt20p interact with each other and with Apg17p, an autophagy-specific protein which interacts with Apg1p, a Ser/Thr protein kinase required for both the Cvt pathway and autophagy (Scott et al. ; Kamada et al. 2000). Interestingly, a third PI3P-binding yeast protein, Etf1p, is also involved in the Cvt pathway (Wurmser and Emr 2002). Etf1p is a type II transmembrane protein which localises to prevacuolar compartments. The in vitro interaction between Etf1p and PI3P requires a basic amino acid motif (KKPAKK) within the cytosolic region of the protein, thus representing a novel PI3P interaction motif distinct from FYVE and PX domains. Deletion of Etf1p or mutation of the KKPAKK motif was found to induce sorting defects in the Cvt pathway. The exact function of Etf1p is presently unknown, but it has been suggested that PI3P functions as a localization determinant of Etf1p. This may serve to bring the protein into proximity with yet-unidentified Etf1p-interacting protein(s), with Apg8p being one suggested candidate. Although three PI3P effectors have been identified as regulators of the Cvt pathways in yeast, no PI3P-binding proteins have so far been identified as specific regulators of autophagy in yeast or mammals. As autophagy requires PI3P (Stromhaug and Klionsky 2001), such proteins are likely to exist.

5
Conclusion

FYVE and PX domains have evolved as protein modules which are essential in eukaryotes. Their ability to bind the low-abundance lipid PI3P, which is specifically formed on endosomes and phagosomes, embodies the cell with a mechanism to recruit a specific subset of cytosolic proteins onto these organelles. Most of the FYVE and PX domain-containing proteins studied so far serve as regulators of membrane traffic, including endocytic membrane fusion, endosomal sorting, endocytic recycling and phagosome maturation. However, others have more specialised functions in signal transduction and microbial killing. Through evolution, the ratio between the numbers of FYVE-containing and PX-containing proteins has remained relatively constant. Why has there been an evolutionary pressure to conserve two distinct domains with comparable affinities for PI3P? The small size of the FYVE domain makes this domain inexpensive to synthesise and easy to accommodate within a given protein sequence. On the other hand, the larger PX domain has the advantage of being suited for protein-protein interactions and is regulatable through phosphorylation and binding to SH3 domains (Simonsen and Stenmark 2001). The two domains may also hold other secrets yet to be revealed, which may further explain why we cannot do without any of them.

Acknowledgements. We thank the Research Council of Norway, the Norwegian Cancer Society, The Novo Nordisk Foundation and the Anders Jahre's Foundation for promotion of Science for financial support.

References

Babior, B.M. (1999). NADPH oxidase: an update. Blood 93, 1464–1476
Barr, V.A., Phillips, S.A., Taylor, S.I., and Haft, C.R. (2000). Overexpression of a novel sorting nexin, SNX15, affects endosome morphology and protein trafficking. Traffic. 1, 904–916
Bilodeau, P.S., Urbanowski, J.L., Winistorfer, S.C., and Piper, R.C. (2002). The Vps27p Hse1p complex binds ubiquitin and mediates endosomal protein sorting. Nat. Cell Biol. 4, 534–539
Bravo, J., Karathanassis, D., Pacold, C.M., Pacold, M.E., Ellson, C.D., Anderson, K.E., Butler, P.J., Lavenir, I., Perisic, O., Hawkins, P.T., Stephens, L., and Williams, R.L.

(2001). The crystal structure of the PX domain from p40(phox) bound to phosphatidylinositol 3-phosphate. Mol. Cell 8, 829–839

Burda, P., Padilla, S.M., Sarkar, S., and Emr, S.D. (2002). Retromer function in endosome-to-Golgi retrograde transport is regulated by the yeast Vps34 PtdIns 3-kinase. J. Cell Sci. 115, 3889–3900

Callaghan, J., Simonsen, A., Gaullier, J.-M., Toh, B.-H., and Stenmark, H. (1999). The endosome fusion regulator EEA1 is a dimer. Biochem. J. 338, 539–543

Ceresa, B.P. and Schmid, S.L. (2000). Regulation of signal transduction by endocytosis. Curr. Opin. Cell Biol. 12, 204–210

Chan, T.O., Rittenhouse, S.E., and Tsichlis, P.N. (1999). AKT/PKB and other D3 phosphoinositide-regulated kinases: kinase activation by phosphoinositide-dependent phosphorylation. Annu. Rev. Biochem. 68:965–1014., 965–1014

Cheever, M.L., Sato, T.K., de Beer, T., Kutateladze, T., Emr, S.D., and Overduin, M. (2001). Phox domain interaction with PtdIns(3)P targets Vam7 t-SNARE to vacuole membranes. Nature Cell Biol. 3, 613–618

Chin, L.-S., Raynor, M.C., Wei, X., Chen, H.-Q., and Li, L. (2001). Hrs interacts with sorting nexin 1 and regulates degradation of epidermal growth factor receptor. J. Biol. Chem. 276, 7069–7078

Christoforidis, S., McBride, H.M., Burgoyne, R.D., and Zerial, M. (1999a). The Rab5 effector EEA1 is a core component of endosome docking. Nature 397, 621–626

Christoforidis, S., Miaczynska, M., Ashman, K., Wilm, M., Zhao, L., Yip, S.-C., Waterfield, M.D., Backer, J.M., and Zerial, M. (1999b). Phosphatidylinositol-3-OH kinases are Rab5 effectors. Nature Cell Biol. 1, 249–252

Cozier, G.E., Carlton, J., McGregor, A.H., Gleeson, P.A., Teasdale, R.D., Mellor, H., and Cullen, P.J. (2002). The phox homology (PX) domain-dependent, 3-phosphoinositide-mediated association of sorting nexin-1 with an early sorting endosomal compartment is required for its ability to regulate epidermal growth factor receptor degradation. J. Biol. Chem. 277, 48730–48736

de Renzis, S., Sonnichsen, B., and Zerial, M. (2002). Divalent Rab effectors regulate the sub-compartmental organization and sorting of early endosomes. Nat. Cell Biol. 4, 124–133

Dumas, J.J., Merithew, E., Sudharshan, E., Rajamani, D., Hayes, S., Corvera, S., and Lambright, D.G. (2001). Structural basis of multivalent phosphatidylinositol 3-phosphate recognition by homodimeric EEA1. Mol. Cell 8, 947–958

Ellson, C.D., Anderson, K.E., Morgan, G., Chilvers, E.R., Lipp, P., Stephens, L.R., and Hawkins, P.T. (2001a). Phosphatidylinositol 3-phosphate is generated in phagosomal membranes. Curr. Biol. 11, 1631–1635

Ellson, C.D., Andrews, S., Stephens, L.R., and Hawkins, P.T. (2002). The PX domain: a new phosphoinositide-binding module. J. Cell Sci. 115, 1099–1105

Ellson, C.D., Gobert-Gosse, S., Anderson, K.E., Davidson, K., Erdjument-Bromage, H., Tempst, P., Thuring, J.W., Cooper, M.A., Lim, Z.-Y., Holmes, A.B., Chilvers, E.R., J., Hawkins, P.T., and Stephens, L.R. (2001b). Phosphatidylinositol 3-phosphate regulates the neutrophil oxidase complex by binding to the PX domain of p40^{phox}. Nature Cell Biol. 3, 679–682

Fratti, R.A., Backer, J.M., Gruenberg, J., Corvera, S., and Deretic, V. (2001b). Role of phosphatidylinositol 3-kinase and Rab5 effectors in phagosomal biogenesis and mycobacterial phagosome maturation arrest. J Cell Biol 154, 631–644

Garin, J., Diez, R., Kieffer, S., Dermine, J.F., Duclos, S., Gagnon, E., Sadoul, R., Rondeau, C., and Desjardins, M. (2001). The phagosome proteome: insight into phagosome functions. J Cell Biol 152, 165–180

Gaullier, J.-M., Rönning, E., Gillooly, D.J., and Stenmark, H. (2000). Interaction of the EEA1 FYVE finger with phosphatidylinositol 3-phosphate and early endosomes. Role of conserved residues. J. Biol. Chem. 275, 24595–24600

Gaullier, J.-M., Simonsen, A., D'Arrigo, A., Bremnes, B., Aasland, R., and Stenmark, H. (1998). FYVE fingers bind PtdIns(3)P. Nature 394, 432–433

Gillooly, D.J., Morrow, I.C., Lindsay, M., Gould, R., Bryant, N.J., Gaullier, J.-M., Parton, R.G., and Stenmark, H. (2000). Localization of phosphatidylinositol 3-phosphate in yeast and mammalian cells. EMBO J. 19, 4577–4588

Gillooly, D.J., Simonsen, A., and Stenmark, H. (2001a). Cellular functions of phosphatidylinositol 3-phosphate and FYVE domain proteins. Biochem. J. 355, 249–258

Gillooly, D.J., Simonsen, A., and Stenmark, H. (2001b). Phosphoinositides and phagocytosis. J. Cell Biol. 155, 15–17

Haft, C.R., de la Luz, S.M., Bafford, R., Lesniak, M.A., Barr, V.A., and Taylor, S.I. (2000). Human orthologs of yeast vacuolar protein sorting proteins Vps26, 29, and 35: assembly into multimeric complexes. Mol Biol Cell 11, 4105–4116

Hayes, S., Chawla, A., and Corvera, S. (2002). TGF beta receptor internalization into EEA1-enriched early endosomes: role in signaling to Smad2. J. Cell Biol. 158, 1239–1249

Hicke, L. (2001). Protein regulation by monoubiquitin. Nat. Rev. Mol. Cell Biol. 2, 195–201

Hiroaki, H., Ago, T., Ito, T., Sumimoto, H., and Kohda, D. (2001). Solution structure of the PX domain, a target of the SH3 domain. Nature Struct. Biol. 8, 526–530

Itoh, F., Divecha, N., Brocks, L., Oomen, L., Janssen, H., Calafat, J., Itoh, S., and Dijke, P.P. (2002). The FYVE domain in Smad anchor for receptor activation (SARA) is sufficient for localization of SARA in early endosomes and regulates TGF-beta/Smad signalling. Genes Cells 7, 321–331

Kamada, Y., Funakoshi, T., Shintani, T., Nagano, K., Ohsumi, M., and Ohsumi, Y. (2000). Tor-mediated induction of autophagy via an Apg1 protein kinase complex. J. Cell Biol. 150, 1507–1513

Kanai, F., Liu, H., Akbary, H., Field, S., Matsuo, T., Brown, G., Cantley, L.C., and Yaffe, M.B. (2001). The PX domains of p47phox and p40phox bind to lipid products of phosphoinositide 3-kinase. Nature Cell Biol. 3, 675–678

Karathanassis, D., Stahelin, R.V., Bravo, J., Perisic, O., Pacold, C.M., Cho, W., and Williams, R.L. (2002). Binding of the PX domain of p47(phox) to phosphatidylinositol 3,4- bisphosphate and phosphatidic acid is masked by an intramolecular interaction. EMBO J. 21, 5057–5068

Kauppi, M., Simonsen, A., Bremnes, B., Vieira, A., Callaghan, J., Stenmark, H., and Olkkonen, V.M. (2002). The small GTPase Rab22 interacts with EEA1 and controls endosomal membrane trafficking. J. Cell Sci. 115, 899–911

Klionsky, D.J., Cueva, R., and Yaver, D.S. (1992). Aminopeptidase I of *Saccharomyces cerevisiae* is localized to the vacuole independent of the secretory pathway. J. Cell Biol. 119, 287–299

Kurten, R.C., Cadena, D.L., and Gill, G.N. (1996). Enhanced degradation of EGF receptors by a sorting nexin, SNX1. Science 272, 1008–1010

Laporte, J., Liaubet, L., Blondeau, F., Tronchere, H., Mandel, J.L., and Payrastre, B. (2002). Functional redundancy in the myotubularin family. Biochem. Biophys. Res. Commun. 291, 305–312

Leber, R., Silles, E., Sandoval, I.V., and Mazon, M.J. (2001). Yol082p, a novel CVT protein involved in the selective targeting of aminopeptidase I to the yeast vacuole. J. Biol. Chem. 276, 29210–29217

Liu, D., Yang, X., and Songyang, Z. (2000). Identification of CISK, a new member of the SGK kinase family that promotes IL-3-dependent survival. Curr Biol 10, 1233–1236

Mao, Y., Nickitenko, A., Duan, X., Lloyd, T.E., Wu, M.N., Bellen, H., and Quiocho, F.A. (2000). Crystal structure of the VHS and FYVE tandem domains of Hrs, a protein involved in membrane trafficking and signal transduction. Cell 100, 447–456

Mari, M., Cormont, M., Mari, S., and Le Marchand-Brustel, Y. (1999). Cloning of Rabip4, an effector of the small GTPase Rab4. Characterisation of its functions in endocytosis. Biochimie 81 (Suppl. 6), s223

Marshall, J.G., Booth, J.W., Stambolic, V., Mak, T., Balla, T., Schreiber, A.D., Meyer, T., and Grinstein, S. (2001). Restricted accumulation of phosphatidylinositol 3-kinase products in a plasmalemmal subdomain during Fc gamma receptor-mediated phagocytosis. J Cell Biol 153, 1369–1380

McBride, H.M., Rybin, V., Murphy, C., Giner, A., Teasdale, R., and Zerial, M. (1999). Oligomeric complexes link Rab5 effectors with NSF and drive membrane fusion via interactions between EEA1 and syntaxin 13. Cell 98, 377–386

Merithew, E., Stone, C., Eathiraj, S., and Lambright, D.G. (2002). Determinants of Rab5 interaction with the N-terminus of early endosome antigen 1. J. Biol. Chem

Misra, S. and Hurley, J.H. (1999). Crystal structure of a phosphatidylinositol 3-phosphate-specific membrane-targeting motif, the FYVE domain of Vps27p. Cell 97, 657–666

Murray, J.T., Panaretou, C., Stenmark, H., Miaczynska, M., and Backer, J.M. (2002). Role of Rab5 in the recruitment of hVps34/p150 to the early endosome. Traffic. 3, 416–427

Nice, D.C., Sato, T.K., Stromhaug, P.E., Emr, S.D., and Klionsky, D.J. (2002). Cooperative binding of the cytoplasm to vacuole targeting pathway proteins, Cvt13 and Cvt20, to phosphatidylinositol 3-phosphate at the pre-autophagosomal structure is required for selective autophagy. J. Biol. Chem. 277, 30198–30207

Nielsen, E., Christoforidis, S., Uttenweiler-Joseph, S., Miaczynska, M., Dewitte, F., Wilm, M., Hoflack, B., and Zerial, M. (2000). Rabenosyn-5, a novel Rab5 effector, is complexed with hVPS45 and recruited to endosomes through a FYVE finger domain. J. Cell Biol. 151, 601–612

Odorizzi, G., Babst, M., and Emr, S.D. (1998). Fab1p PtdIns(3)P 5-kinase function essential for protein sorting in the multivesicular body. Cell 95, 847–858

Panopoulou, E., Gillooly, D.J., Wrana, J.L., Zerial, M., Stenmark, H., Murphy, C., and Fotsis, T. (2002). Early endosomal regulation of Smad-dependent signaling in endothelial cells. J. Biol. Chem. 277, 18046–18052

Pattni, K., Jepson, M., Stenmark, H., and Banting, G. (2001). A PtdIns(3)P-specific probe cycles on and off host cell membranes during *Salmonella* invasion of mammalian cells. Curr. Biol. 11, 1636–1642

Peterson, M.R., Burd, C.G., and Emr, S.D. (1999). Vac1p coordinates rab and phosphatidylinositol 3-kinase signaling in Vps45p-dependent vesicle docking/fusion at the endosome. Curr. Biol. 9, 159–162

Ponting, C.P. (1996). Novel domains in NADPH oxidase subunits, sorting nexins, and PtdIns 3-kinases: binding partners of SH3 domains? Protein Sci. 5, 2353–2357

Raiborg, C., Bache, K.G., Gillooly, D.J., Madshus, I.H., Stang, E., and Stenmark, H. (2002). Hrs sorts ubiquitinated proteins into clathrin-coated microdomains of early endosomes. Nature Cell Biol. 4, 394–398

Raiborg, C., Bache, K.G., Mehlum, A., Stang, E., and Stenmark, H. (2001a). Hrs recruits clathrin to early endosomes. EMBO J. 20, 5008–5021

Raiborg, C., Bremnes, B., Mehlum, A., Gillooly, D.J., Stang, E., and Stenmark, H. (2001b). FYVE and coiled-coil domains determine the specific localisation of Hrs to early endosomes. J. Cell Sci. 114, 2255–2263

Raiborg, C. and Stenmark, H. (2002). Hrs and endocytic sorting of ubiquitinated membrane proteins. Cell Struct. Funct. 27, 439–444

Roggo, L., Bernard, V., Kovacs, A.L., Rose, A.M., Savoy, F., Zetka, M., Wymann, M.P., and Muller, F. (2002). Membrane transport in *Caenorhabditis elegans*: an essential role for VPS34 at the nuclear membrane. EMBO J. 21, 1673–1683

Rubino, M., Miaczynska, M., Lippé, R., and Zerial, M. (2000). Selective membrane recruitment of EEA1 suggests a role in directional transport of clathrin-coated vesicles to early endosomes. J. Biol. Chem. 275, 3745–3748

Sachse, M., Urbe, S., Oorschot, V., Strous, G.J., and Klumperman, J. (2002). Bilayered clathrin coats on endosomal vacuoles are involved in protein sorting toward lysosomes. Mol. Biol. Cell 13, 1313–1328

Sankaran, V.G., Klein, D.E., Sachdeva, M.M., and Lemmon, M.A. (2001). High-affinity binding of a FYVE domain to phosphatidylinositol 3-phosphate requires intact phospholipid but not FYVE domain oligomerization. Biochemistry 40, 8581–8587

Sato, T.K., Overduin, M., and Emr, S.D. (2001). Location, location, location: membrane targeting directed by PX domains. Science 294, 1881–1885

Sbrissa, D., Ikonomov, O.C., and Shisheva, A. (1999). PIKfyve, a mammalian ortholog of yeast Fab1p lipid kinase, synthesizes 5-phosphoinositides. J. Biol. Chem. 274, 21589–21597

Schu, P.V., Takegawa, K., Fry, M.J., Stack, J.H., Waterfield, M.D., and Emr, S.D. (1993). Phosphatidylinositol 3-kinase encoded by yeast VPS34 gene essential for protein sorting. Science 260, 88–91

Schwarz, D.G., Griffin, C.T., Schneider, E.A., Yee, D., and Magnuson, T. (2002). Genetic analysis of sorting nexins 1 and 2 reveals a redundant and essential function in mice. Mol. Biol. Cell 13, 3588–3600

Scott, S.V., Guan, J., Hutchins, M.U., Kim, J., and Klionsky, D.J. (2001). Cvt19 is a receptor for the cytoplasm-to-vacuole targeting pathway. Mol. Cell 7, 1131–1141

Scott, S.V. and Klionsky, D.J. (1998). Delivery of proteins and organelles to the vacuole from the cytoplasm. Curr. Op

Scott, S.V., Nice, D.C., III, Nau, J.J., Weisman, L.S., Kamada, Y., Keizer-Gunnink, I., Funakoshi, T., Veenhuis, M., Ohsumi, Y., and Klionsky, D.J. Apg13p and Vac8p are part of a complex of phosphoproteins that are required for cytoplasm to vacuole targeting

Seaman, M.N.J., McCaffery, J.M., and Emr, S.D. (1998). A membrane coat complex essential for endosome-to-Golgi retrograde transport in yeast. J. Cell Biol. 142, 665–681

Shih, S.C., Katzmann, D.J., Schnell, J.D., Sutanto, M., Emr, D., and Hicke, L. (2002). Epsins and Vps27p/Hrs contain ubiquitin-binding domains that function in receptor endocytosis. Nat. Cell Biol. 4, 389–393

Shintani, T., Huang, W.P., Stromhaug, E., and Klionsky, D.J. (2002). Mechanism of cargo selection in the cytoplasm to vacuole targeting pathway. Dev. Cell 3, 825–837

Simonsen, A., Gaullier, J.-M., D'Arrigo, A., and Stenmark, H. (1999). The Rab5 effector EEA1 interacts directly with syntaxin-6. J. Biol. Chem. 274, 28857–28860

Simonsen, A., Lippé, R., Christoforidis, S., Gaullier, J.-M., Brech, A., Callaghan, J., Toh, B.-H., Murphy, C., Zerial, M., and Stenmark, H. (1998). EEA1 links PI(3)K function to Rab5 regulation of endosome fusion. Nature 394, 494–498

Simonsen, A. and Stenmark, H. (2001). PX domains: attracted by phosphoinositides. Nat. Cell Biol. 3, E179-E182

Simonsen, A., Wurmser, A.E., Emr, S.D., and Stenmark, H. (2001). The role of phosphoinositides in membrane transport. Curr. Opin. Cell Biol. 13, 485–492

Song, X., Zhang, A., Huang, G., Liang, X., Virbasius, J.V., Czech, M.P., and Zhou, G.W. (2001). Phox homology (PX) domains specifically bind phosphatidylinositol phosphates. Biochemistry 40, 8940–8944

Sorkin, A. and Von Zastrow, M. (2002). Signal transduction and endocytosis: close encounters of many kinds. Nat. Rev. Mol. Cell Biol. 3, 600–614

Stack, J.H. and Emr, S.D. (1994). Vps34p required for yeast vacuolar protein sorting is a multiple specificity kinase that exhibits both protein kinase and phosphatidylinositol-specific PI 3-kinase activities. J. Biol. Chem. 269, 31552–31562

Stenmark, H., Aasland, R., and Driscoll, P.C. (2002). The phosphatidylinositol 3-phosphate-binding FYVE finger. FEBS Lett. 513, 77–84

Stenmark, H., Aasland, R., Toh, B.H., and D'Arrigo, A. (1996). Endosomal localization of the autoantigen EEA1 is mediated by a zinc-binding FYVE finger. J. Biol. Chem. 271, 24048–24054

Stenmark, H. and Gillooly, D.J. (2001). Intracellular trafficking and turnover of phosphatidylinositol 3-phosphate. Semin. Cell Dev. Biol. 12, 193–199

Stromhaug, P.E. and Klionsky, D.J. (2001). Approaching the molecular mechanism of autophagy. Traffic. 2, 524–531

Tall, G.G., Hama, H., DeWald, D.B., and Horazdovsky, B.F. (1999). The phosphatidylinositol 3-phosphate binding protein Vac1p interacts with a Rab GTPase and a Sec1p homologue to facilitate vesicle-mediated vacuolar protein sorting. Mol. Biol. Cell 10, 1873–1889

Taylor, G.S., Maehama, T., and Dixon, J.E. (2000). Myotubularin, a protein tyrosine phosphatase mutated in myotubular myopathy, dephosphorylates the lipid second messenger, phosphatidylinositol 3-phosphate. Proc. Natl. Acad. Sci. USA 97, 8910–8915

Teasdale, R.D., Loci, D., Houghton, F., Karlsson, L., and Gleeson, P.A. (2001). A large family of endosome-localized proteins related to sorting nexin 1. Biochem. J. 358, 7–16

Tsukazaki, T., Chiang, T.A., Davison, A.F., Attisano, L., and Wrana, J.L. (1998). SARA, a FYVE domain protein that recruits Smad2 to the TGFβ receptor. Cell 95, 779–791

Urbé, S., Mills, I.G., Stenmark, H., Kitamura, N., and Clague, M.J. (2000). Endosomal localisation and receptor dynamics determine tyrosine phosphorylation of hepatocyte growth factor-regulated tyrosine kinase substrate. Mol. Cell. Biol. 20, 7685–7692

Vanhaesebroeck, B. and Waterfield, M.D. (1999). Signaling by distinct classes of phosphoinositide 3-kinases. Exp. Cell Res. 253, 239–254

Vieira, O.V., Botelho, R.J., Rameh, L., Brachmann, S.M., Matsuo, T., Davidson, H.W., Schreiber, A., Backer, J.M., Cantley, L.C., and Grinstein, S. (2001). Distinct roles of class I and III phosphatidylinositol 3-kinases in phagosome formation and maturation. J. Cell Biol. 155, 19–25

Vignais, P.V. (2002). The superoxide-generating NADPH oxidase: structural aspects and activation mechanism. Cell Mol. Life Sci. 59, 1428–1459

Virbasius, J.V., Song, X., Pomerleau, D.P., Zhan, Y., Zhou, G.W., and Czech, M.P. (2001). Activation of the Akt-related cytokine-independent survival kinase requires interaction of its phox domain with endosomal phosphatidylinositol 3-phosphate. Proc. Natl. Acad. Sci. U. S. A 98, 12908–12913

Walker, D.M., Urbé, S., Dove, S.K., Tenza, D., Raposo, G., and Clague, M.J. (2001). Characterization of MTMR3: an inositol lipid 3-phosphatase with novel substrate activity. Curr. Biol. 11, 1600–1605

Wishart, M.J. and Dixon, J.E. (2002). PTEN and myotubularin phosphatases: from 3-phosphoinositide dephosphorylation to disease. Phosphatase and tensin homolog deleted on chromosome ten. Trends Cell Biol. 12, 579–585

Wishart, M.J., Taylor, G.S., and Dixon, J.E. (2001). Phoxy lipids: revealing PX domains as phosphoinositide binding modules. Cell 105, 817–820

Worby, C.A. and Dixon, J.E. (2002). Sorting out the cellular functions of sorting nexins. Nat. Rev. Mol. Cell Biol. 3, 919–931

Wurmser, A.E. and Emr, S.D. (1998). Phosphoinositide signaling and turnover: PtdIns(3)P, a regulator of membrane traffic, is transported to the vacuole and degraded by a process that requires lumenal vacuolar hydrolase activities. EMBO J. 17, 4930–4942

Wurmser, A.E. and Emr, S.D. (2002). Novel PtdIns(3)P-binding protein Etf1 functions as an effector of the Vps34 PtdIns 3-kinase in autophagy. J. Cell Biol. 158, 761–772

Xu, Y.H.H., Seet, L., Wong, S.H., and Hong, W. (2001). Sorting nexin 3 (SNX3) regulates endosomal function via its PX domain-mediated interaction with phosphatidylinositol 3-phosphate. Nature Cell Biol. 3, 658–666

Yu, J.W. and Lemmon, M.A. (2001). All phox homology (PX) domains from *Saccharomyces cerevisiae* specifically recognize phosphatidylinositol 3-phosphate. J. Biol. Chem. 276, 44179–44184

Regulation of the Actin Cytoskeleton by PI(4,5)P$_2$ and PI(3,4,5)P$_3$

P. Hilpelä · M. K. Vartiainen · P. Lappalainen

Program in Cellular Biotechnology, Institute of Biotechnology,
University of Helsinki, P.O. Box 56, 00014 Helsinki, Finland
E-mail: pekka.lappalainen@helsinki.fi

1	The Actin Cytoskeleton	118
2	Regulation of Actin Dynamics by Phosphoinositides In Vivo	121
2.1	General Features of Phosphoinositides	121
2.2	Regulation of the Actin Cytoskeleton by PI(4,5)P$_2$	123
2.3	Regulation of the Actin Cytoskeleton by PI(3,4,5)P$_3$	125
2.4	Regulation of the Actin Cytoskeleton by Other Phosphoinositides	126
3	Interactions of Actin-Associated Proteins with Phosphatidylinositol Phosphates	127
3.1	Actin Filament-Capping and -Severing Proteins	127
3.1.1	Capping Protein	127
3.1.2	Gelsolin Family Proteins	128
3.2	Actin Monomer-Binding Proteins	130
3.2.1	Profilin	130
3.2.2	ADF/Cofilin	132
3.2.3	Twinfilin	134
3.3	Proteins That Promote Actin Filament Nucleation	134
3.3.1	WASP/N-WASP	134
3.3.2	Cortactin	135
3.4	Actin Cross-Linking Proteins	137
3.4.1	α-Actinin	137
3.4.2	Filamin	138
3.4.3	Cortexillin	139
3.5	Proteins Linking the Actin Cytoskeleton to the Plasma Membrane	139
3.5.1	ERM Proteins	139
3.5.2	Talin	141
3.5.3	Vinculin	141
3.6	Phosphoinositide-Responsive Activators of the Rho Family GTPases	142
3.7	Others	144
3.7.1	GMC Proteins	144
3.7.2	Myosin X	146
4	The Molecular Mechanism of Phosphoinositide-Promoted Actin Assembly	147
	References	149

Abstract The actin cytoskeleton is fundamental for various motile and morphogenetic processes in cells. The structure and dynamics of the actin cytoskeleton are regulated by a wide array of actin-binding proteins, whose activities are controlled by various signal transduction pathways. Recent studies have shown that certain membrane phospholipids, especially $PI(4,5)P_2$ and $PI(3,4,5)P_3$, regulate actin filament assembly in cells and in cell extracts. $PI(4,5)P_2$ appears to be a general regulator of actin polymerization at the plasma membrane or at membrane microdomains, whereas $PI(3,4,5)P_3$ promotes the assembly of specialized actin filament structures in response to some growth factors. Biochemical studies have demonstrated that the activities of many proteins promoting actin assembly are upregulated by $PI(4,5)P_2$, whereas proteins that inhibit actin assembly or promote filament disassembly are downregulated by $PI(4,5)P_2$. $PI(3,4,5)P_3$ promotes its effects on the actin cytoskeleton mainly through activation of the Rho family of small GTPases. In addition to their effects on actin dynamics, both $PI(4,5)P_2$ and $PI(3,4,5)P_3$ promote the formation of specific actin filament structures through activation/inactivation of actin filament cross-linking proteins and proteins that mediate cytoskeleton-plasma membrane interactions.

1
The Actin Cytoskeleton

Actin is a highly conserved protein that is essential for a variety of cellular processes in eukaryotes. In muscles, actin filaments are relatively stable and are organized into an almost crystalline array. Together with myosin filaments they are responsible for contracting muscle and generating force. In other cell types, actin filaments are highly dynamic and are central to many cellular processes, including motility, morphogenesis, endocytosis, secretion, and cytokinesis (Sheterline et al. 1998). Actin is also found in the nucleus, where it may be involved in chromatin remodeling and transcriptional control (Olave et al. 2002). Probably the best-understood actin-driven process in non-muscle cells is motility. In migrating cells a specialized network of actin filaments assembles at their leading edges and promotes pseudopod extension and cell protrusion. The network is polar, with actin's plus-ends near the plasma membrane and minus-ends away from the leading edge. Force is produced by coordinated assembly at the plus-ends and disassembly at the minus-ends (Pollard et al. 2000).

Actin is a ~43-kDa globular protein that exists in two forms: G-actin (globular) is the monomeric form that can assemble into the polar filamentous F-actin (microfilaments). A nucleotide, either ATP or ADP, is bound in a cleft that separates the two lobes of actin and is associated with a divalent cation. The nucleotide status (ATP- or ADP bound) profoundly impacts the biochemical properties of actin: ATP-actin monomers have a higher affinity for filament ends and thus more readily assemble into filaments. Filamentous actin is an ATPase, so assembly activates ATP hydrolysis and the polar filaments form two chemically distinct ends: The plus-end (barbed end) contains ATP-actin, is fast growing, and assembles at steady state; the minus-end (pointed end) contains ADP subunits, is slow growing, and disassembles at steady state (Sheterline et al. 1998).

The structure and dynamics of the actin cytoskeleton are regulated by an array of actin-binding proteins that interact with monomers or filaments. The Arp2/3 complex and formin family proteins accelerate the formation of new filaments and regulate actin filament nucleation at cell regions where new assembly is required (Welch and Mullins 2002; Pruyne et al. 2002; Sagot et al. 2002a, b). The Arp2/3 complex also cross-links actin filaments to form the specialized networks that are in the leading edges of migrating cells (Mullins et al. 1998). Also, other actin cross-linking proteins such as α-actinin and filamin regulate the formation of actin filament bundles and networks in specialized cell regions (Stossel et al. 2001). Furthermore, some myosins cross-link actin filaments and promote filament contractility, whereas other myosins are involved in the transport of cargo along actin filaments (Sellers 2000).

Cellular actin filaments are dynamic, so turnover is frequent and necessary. Turnover is aided by proteins that sever or depolymerize actin filaments. The ADF/cofilins promote rapid actin dynamics by depolymerizing filaments at their minus-ends (Carlier et al. 1999), whereas gelsolins sever actin filaments and cap the filaments' plus-ends in a calcium-responsive manner (Weeds and Maciver 1993; Sun et al. 1999). In addition to the gelsolins, heterodimeric capping proteins bind to the plus-ends of the filaments and prevent assembly (Cooper and Schafer 2000). The assembly of actin monomers into filament ends is also regulated by a number of proteins that bind actin monomers such as profilin, twinfilin, and thymosin-β4. Each of these proteins has distinct effects on the dynamics and localization of the actin monomer pool and thus regulates different aspects of actin dynamics (Pollard et al. 2000; Palmgren et al. 2002).

The activities of actin-binding proteins are regulatedby various signaling pathways. This regulation allows for the formation of specific actin structures at certain regions in a cell. The most thoroughly characterized regulators of actin-binding proteins are the small GTPases of the Rho family. They can be divided into three subfamilies: Rho, Rac, and Cdc42. Each of these proteins mediates diverse signals (e.g., from growth factor receptors) to the actin cytoskeleton. Rho family GTPases regulate actin dynamics either by directly interacting with regulators of actin dynamics or by regulating the activities of these proteins through phosphorylation pathways. An example of direct binding is the activation of Arp2/3-mediated actin nucleation by Cdc42 and Rac; Cdc42 directly interacts with and activates the large multifunctional actin-binding protein Wiskott-Aldrich Syndrome protein (WASP), which consequently activates the Arp2/3 complex (Rohatgi et al. 1999). Similarly, activated Rac induces actin nucleation by disassembling the inactive Scar protein complex, which then activates the Arp2/3 complex with a mechanism similar to that of activated WASP (Machesky and Insall 1998; Eden et al. 2002).

Rho family GTPases also activate or inactivate certain actin-binding proteins by kinase/phosphatase pathways. Rho downregulates the activity of ADF/cofilin protein through phosphorylation by ROCK and LIM kinases (Maekawa et al. 1999). ROCK kinase also phosphorylates certain myosins and inactivates myosin light chain phosphatase, thereby resulting in an increase in myosin phosphorylation and an increase in actomyosin contraction (Bresnick 1999).

In addition to regulation by GTPases and protein phosphorylation/dephosphorylation pathways, the dynamics of the actin cytoskeleton are regulated by phosphorylated phosphatidylinositols. Although these membrane phospholipids were classically considered as secondary messengers, recent evidence has shown that they also directly regulate various cellular functions and structures, including the actin cytoskeleton. In this review we discuss the in vivo roles of the phosphatidylinositol phosphates $PI(4,5)P_2$ and $PI(3,4,5)P_3$ in actin dynamics and discuss their interactions with and effects on various actin-binding proteins.

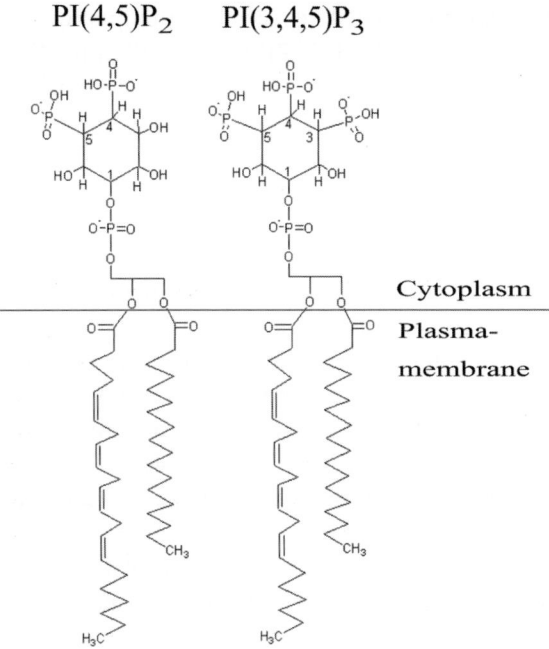

Fig. 1 Chemical structures of PI(4,5)P$_2$ and PI(3,4,5)P$_3$

2
Regulation of Actin Dynamics by Phosphoinositides In Vivo

A number of recent studies have revealed an important role for phosphoinositide signaling in the regulation of the actin cytoskeleton. The various phosphorylated phosphoinositides control the activities and localizations of numerous actin-binding proteins and of other proteins that regulate the actin cytoskeleton.

2.1
General Features of Phosphoinositides

Phosphoinositide is a collective term for phosphatidylinositol (PI) and its phosphorylated derivatives (Fruman et al. 1998; Martin 1998). PI consists of d-*myo*-inositol-1-phosphate linked via its phosphate group to diacylglycerol. The d-myo-inositol headgroup of PI contains five hydroxyl groups, three of which are targets of phosphorylation (Fig. 1). These positions may be reversibly phosphorylated in various combinations to yield a number of PI derivatives: PI(3)P, PI(4)P, PI(5)P, PI(3,4)P_2, PI(3,5)P_2, PI(4,5)P_2, and PI(3,4,5)P_3. Among the eight phosphoinositide species present in eukaryotic cells, PI is the most abundant. Approximately 5% of cellular PI is phosphorylated at position D-4 [PI(4)P], and another 5% is phosphorylated at both positions D-4 and D-5 [PI(4,5)P_2] (Rameh and Cantley 1999).

PI(4,5)P_2 comprises about 1% of the total phospholipid content in the plasma membrane, an effective concentration of approximately 10 μM (McLaughlin et al. 2002). PI(4,5)P_2 is also found in the Golgi compartment (Jones et al. 2000), endosomes, endoplasmic reticulum, and electron-dense structures within the nucleus (Watt et al. 2002). PI(4,5)P_2 is mainly generated by phosphorylating PI(4)P at the D-5 position of the inositol ring by PI(4)P 5-kinase (Ishihara et al. 1998). PI(4)P 5-kinase is a type I PIP kinase and has three subtypes: Iα, Iβ, and Iγ (Ishihara et al. 1996; Loijens and Anderson 1996; Ishihara et al. 1998). Some small GTPases that regulate the actin cytoskeleton do activate the PI(4)P 5-kinases. Rac directly interacts with and activates PI(4)P 5-kinase, and this increases the number of free actin filament plus-ends (Hartwig et al. 1995; Tolias and Carpenter 2000; Tolias et al. 2000). PI(4)P 5-kinase is also an effector of Rho (Chong et al. 1994; Ren and Schwartz 1998), and the two directly interact (Ren et al. 1996). Activation of PI(4)P 5-kinase via Rho triggers ezrin/radixin/moesin (ERM) and induces microvilli formation (Matsui et al. 1999). Another small GTPase, ARF6, activates PI(4)P 5-kinase and thereby promotes membrane ruffling (Honda et al. 1999).

The other phosphoinositides are present at much lower levels in cells, and less than 0.25% of the total inositol-containing lipids are phosphorylated at position D-3 (Rameh and Cantley 1999). Even after receptor stimulation, the cellular concentration of PI(3,4,5)P_3 is at least 25 times lower than that of PI(4,5)P_2 (Lemmon and Ferguson 2000). Monophosphorylated PI(3)P is constitutively present in mammalian cells (Auger et al. 1989) and is thought to play a general role in metabolism, whereas the more highly phosphorylated PIPs function as stimulus-responsive secondary messengers (Auger et al. 1989; Varticovski et al. 1989, 1991; Serunian et al. 1990;). In resting cells, both PI(3,4)P_2 and PI(3,4,5)P_3 are practically undetectable (Auger et al. 1989). After receptor stimulation, there is transient production of PI(3,4,5)P_3 by class I PI 3-kinases and, after a short lag, dephosphorylation of PI(3,4,5)P_3 to PI(3,4)P_2 by inositol polyphosphate-5-phosphatases (PIP-5-phosphatases) (Stephens et al.

1991; Auger and Cantley 1991). Consistent with this, PI(3,4,5)P$_3$ accumulation is immediate and transient, whereas PI(3,4)P$_2$ accumulation is delayed and significantly more sustained (Stephens et al. 1993; Franke et al. 1997). PI(3,4,5)P$_3$ is enriched at the plasma membrane (Gray et al. 1999; Czech 2000).

2.2
Regulation of the Actin Cytoskeleton by PI(4,5)P$_2$

For many years it was assumed that PI(4,5)P$_2$ was solely a precursor to inositol 1,4,5-trisphosphate (IP$_3$) and diacylglycerol (DAG). Although this is a major function of PI(4,5)P$_2$ in cells, evidence has accumulated for the fact that this lipid has other important cellular functions. First, PI(4,5)P$_2$ is a substrate for phosphoinositide kinases that produce PI(3,4,5)P$_3$, and second, PI(4,5)P$_2$ directly interacts with intracellular proteins, including a large number of regulators of the actin cytoskeleton.

The induction and maintenance of a polarized distribution of different actin arrays is dependent on extracellular and intracellular signals. Inositol lipids affect actin indirectly, by interacting with a variety of actin-binding proteins and either stimulating or inhibiting their effects on actin assembly. Actin-binding proteins are the most commonly affected by PI(4,5)P$_2$, but the other inositol lipids, especially PI(3,4,5)P$_3$, are also important in regulation of actin cytoskeleton.

Vesicles containing PI(4,5)P$_2$ or PI(3,4,5)P$_3$ induce actin assembly in *Xenopus* extracts, suggesting that these phospholipids can regulate actin dynamics in vivo (Ma et al. 1998). The cellular roles of phosphoinositides have been also investigated by studying cells with elevated PI(4,5)P$_2$ levels. Overexpressing type I PI(4)P 5-kinase dramatically increases actin polymerization in many cell types. However, these phenotypes do vary between different cell-types (Fig. 2): CV1 cells develop robust stress fibers and form fewer ruffles in response to PDGF (Yamamoto et al. 2001); Swiss 3T3 and REF52 cells form cometlike actin tails and actin-rich spots (Rozelle et al. 2000); NIH 3T3 cells form microvilli (Matsui et al. 1999), and COS-7 cells form pine needle-like actin filament structures (Shibasaki et al. 1997). The role of PI(4,5)P$_2$ in actin assembly is also supported by the observations that reduction of PI(4,5)P$_2$ levels, either through sequestration by overexpressing the PLCδ PH domain or PI(4,5)P$_2$–binding peptide or by overexpressing the PI(4,5)P$_2$ 5-phos-

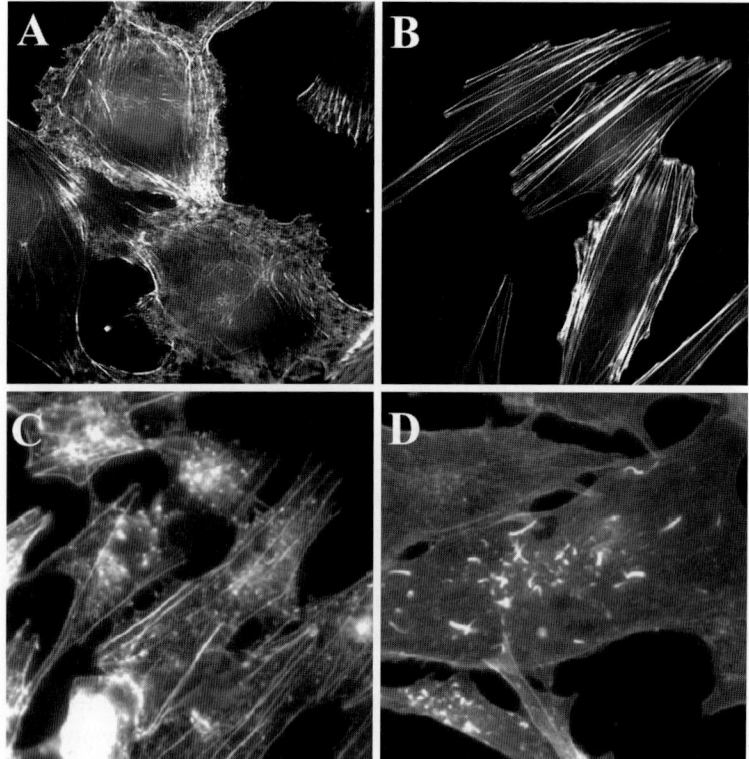

Fig. 2A–D Effects of elevated PI(4,5)P$_2$ levels on the actin cytoskeleton in CV1 and HeLa cells. **A** CV1 control cells. **B** CV1 cells overexpressing PI(4)P 5-kinase. **C** HeLa control cells. **D** HeLa cells overexpressing PI(4)P 5-kinase. An increase in the cellular PI(4,5)P$_2$ level in CV1 cells promotes robust stress fiber formation and inhibits membrane ruffling. In contrast, in HeLa cells the PI(4)P 5-kinase overexpression induces actin comets. (The cell images were kindly provided by Dr. Helen Yin, University of Texas, Southwestern Medical Center)

phatase, inhibits actin polymerization in COS-7 cells (Sakisaka et al. 1997), in NIH 3T3 cells (Raucher et al. 2000), in platelets (Hartwig et al. 1995), and in neutrophils (Glogauer et al. 2000).

Although PI(4,5)P$_2$ clearly regulates actin assembly in vivo, the PI(4,5)P$_2$ levels in most cell types are relatively constant and thus do not change enough to cause a major shift in cytoskeletal behavior. It has been proposed that PI(4,5)P$_2$ accumulates at plasmalemmal lipidmicrodomains, where it modulates the activity of proteins that regulate the actin cytoskeleton. Concentration of PI(4,5)P$_2$ in lipid rafts would provide a permissive environment for tight spatial control of actin assembly; however, the existence of PI(4,5)P$_2$ rafts is still unclear, because studies with different methods are contradictory. The evidence that supports the presence of PI(4,5)P$_2$ in rafts is that PI(4,5)P$_2$ localizes in clusters on the plasma membrane when visualized by anti-PI(4,5)P$_2$ antibodies (Laux et al. 2000), that the PI(4)P 5-kinases are recruited to rafts by the small GTPases Rho and Arf and activate local PI(4,5)P$_2$ synthesis (Chong et al. 1994; Honda et al. 1999), that the GMC proteins (GAP43, MARCKS, and CAP23) directly bind to PI(4,5)P$_2$ and promote its retention in rafts (Laux et al. 2000), and finally that two PI(4)P 5-kinase isoforms accumulate in focal contacts by interacting with talin (Ling et al. 2002; Di Paolo et al. 2002). The evidence against the presence of PI(4,5)P$_2$ in rafts is that the PH domain of PLCδ, which specifically interacts with PI(4,5)P$_2$, is uniformly distributed in the plasma membrane (Stauffer et al. 1998) or in membrane ruffles (Honda et al. 1999; Tall et al. 2000), in agreement with electron micrographs of ultrathin sections showing that the concentration of PI(4,5)P$_2$ is only elevated in lamellipodia and not in microdomains of the plasma membrane (Watt et al. 2002). It was also recently shown that increased levels of GFP-labeled PH domains in plasma membrane patches (including ruffles and lamellipodia) colocalize with various lipophilic membrane dyes, suggesting a general increase in lipid content rather than a specific PI(4,5)P$_2$ enrichment (Van Rheenen and Jalink 2002).

Because of these contrary results, it is unclear how PI(4,5)P$_2$ signals to the actin cytoskeleton. An attractive hypothesis is that PI(4,5)P$_2$ does not act as a stimulus-responsive instructive signal but rather provides a spatial marker to direct actin polymerization close to plasma membrane (Insall and Weiner 2001). This would ensure that actin filaments grow in the correct direction by orienting the barbed ends of new filaments toward the cell surface.

2.3
Regulation of the Actin Cytoskeleton by PI(3,4,5)P$_3$

Compared with PI(4,5)P$_2$, the evidence is clear that PI(3,4,5)P$_3$ levels are well controlled both spatially and temporally. In comparison to the static plasma membrane localization of the PH domains specific for PI(4,5)P$_2$, the PI(3,4,5)P$_3$-specific PH domains rapidly (in ~10 s) accumulate at the plasma membrane in response to specific signals (Stephens et al. 1991). PI 3-kinase phosphorylates $PI(4,5)P_2$ to generate $PI(3,4,5)P_3$ and is activated by certain growth factor receptors at the plasma membrane. Once produced, $PI(3,4,5)P_3$ is dephosphorylated in few minutes by 3-phosphatases such as PTEN (Maehama et al. 2001) or by 5-phosphatases such as SHIP1 or SHIP2 (Lioubin et al. 1996; Damen et al. 1996; Ishihara et al. 1999). PTEN yields $PI(4,5)P_2$, and SHIP1/SHIP2 yields $PI(3,4)P_2$. This fast on-and-off regulation at the plasma membrane strictly regulates signaling in space and time.

PI 3-kinase activity is necessary and sufficient for many processes involving cytoskeletal rearrangements. Constitutively active PI 3-kinase induces a three- to fourfold increase in the motility and invasiveness of epithelial cells across collagen matrices (Keely et al. 1997) and a twofold increase in carcinoma cells (Shaw et al. 1997). Furthermore, direct delivery of $PI(3,4,5)P_3$ to cells causes increased actin polymerization and migration (Derman et al. 1997; Niggli 2000). The specific inhibitors of PI 3-kinases, wortmannin and LY294002, inhibit the motility of MDCK cells (Sander et al. 1998) and carcinoma cells (Shaw et al. 1997). Inhibition of PI 3-kinase by wortmannin in epithelial cells expressing constitutively active Cdc42 and Rac results in a rearrangement of the actin cytoskeleton and an abolishment of lamellipodia and stress fibers (Keely et al. 1997). $PI(3,4,5)P_3$ also seems to control cell polarity during G protein- or PDGF-mediated chemotaxis (Wennström et al. 1994a, b; Wymann and Arcaro 1994; Hawkins et al. 1995; Meili et al. 1999; Hirsch et al. 2000; Li et al. 2000; Sasaki et al. 2000; Servant et al. 2000; Haugh et al. 2000). Activation of the uniformly distributed G protein-coupled receptors on the cell surface leads to the intracellular stimulation of PI 3-kinases and a highly polarized accumulation of $PI(3,4,5)P_3$ and/or $PI(3,4)P_2$ in the plasma membrane at the leading edges of polarized and migrating cells (Meili et al. 1999; Servant et al. 2000; Rickert et al. 2000; Chung et al. 2001). Thus spatial distribution of $PI(3,4,5)P_3$ visualized by PH domains closely matches that of actin polymerization.

2.4
Regulation of the Actin Cytoskeleton by Other Phosphoinositides

There is little evidence that phosphoinositides other than $PI(4,5)P_2$ and $PI(3,4,5)P_3$ regulate actin assembly. It is possible that $PI(3,4)P_2$, a product of $PI(3,4,5)P_3$'s downregulation, has a specific role in regulating the actin cytoskeleton, because two actin-associated proteins, TAPP1 and TAPP2, do specifically bind to PI(3,4)P$_2$ (Dowler et al. 2000; Thomas et al. 2001; Marshall et al. 2002). Based on the sustained presence of PI(3,4)P$_2$ after PI(3,4,5)P$_3$ dephosphorylation, the PI(3,4)P$_2$ pool could activate proteins required for the processes that follow fast actin assembly. A possible candidate for a PI(3,4)P$_2$ regulated protein represents SWAP-70 that has been proposed to prevent premature bundling of newly assembled actin filaments in the lamella behind the leading edge (Hilpelä et al. 2003).

3
Interactions of Actin-Associated Proteins with Phosphatidylinositol Phosphates

The actin-associated proteins bind phosphoinositides with sites that vary from short sequence motifs to well-organized and conserved protein domains, such as the pleckstrin homology (PH) domain. These binding modules differ in their affinities, stoichiometries, and specificities for different phosphoinositides. Some proteins are activated by phosphoinositide binding, whereas others are inactivated.

3.1
Actin Filament-Capping and -Severing Proteins

Proteins that cap and sever actin filaments inhibit assembly and promote depolymerization. PI(4,5)P$_2$ downregulates the activities of capping proteins and the gelsolin family of F-actin-severing proteins, suggesting that phosphoinositides inhibit the capping and severing of actin filaments and promote the formation of long, stable filaments.

3.1.1
Capping Protein

Capping protein is a ubiquitous, heterodimeric protein that binds to the plus-end of actin filaments and prevents actin monomers from exchanging with the filament ends. Capping protein is a central regulator of actin dynamics in vivo, because mutations in genes encoding capping protein subunits in the budding yeast and *Drosophila* disrupt the actin cytoskeleton (Amatruda et al. 1990; Hopmann et al. 1996). In striated muscle cells, a muscle-specific isoform localizes to the Z-disk and attaches the barbed ends of actin filaments to the Z-line (Schafer et al. 1994; Hart and Cooper 1999).

$PI(4,5)P_2$ is the only known regulator of capping protein. $PI(4,5)P_2$ rapidly and efficiently dissociates capping protein from the plus-end of filaments and does so in a concentration-dependent manner. This uncapping process is relatively specific for $PI(4,5)P_2$, because $PI(4)P$ and $PI(3,4,5)P_3$ do not significantly affect capping protein's activity (Schafer et al. 1996). The phosphoinositide binding site on capping protein is unknown.

Uncapping of filament plus-ends by $PI(4,5)P_2$ occurs during platelet activation. In resting platelets, the filament plus-ends are inaccessible to polymerization because of capping proteins. Exposing platelets to agonists such as thrombin initiates a signal transduction cascade that leads to a $PI(4,5)P_2$-dependent uncapping of filament plus-ends and subsequent actin filament assembly. Both the heterodimeric capping protein and gelsolin (see below) coordinately regulate filament plus-end capping in platelets (Barkalow et al. 1996).

3.1.2
Gelsolin Family Proteins

Gelsolin is the founding member of a family of proteins that cap and sever actin filaments. This family has several members including gelsolin, villin, adseverin, CapG, advillin, and supervillin. These proteins share three to six copies of a homologous, ancestral domain. Villin, advillin, and supervillin have other domains, too, and they have specialized roles in regulating the organization of actin filaments in restricted cell types. $PI(4,5)P_2$ appears to inhibit the activity of all gelsolin family proteins (Weeds and Maciver 1993; Sun et al. 1999).

Gelsolin is a Ca^{2+}- and $PI(4,5)P_2$-regulated protein that severs and caps actin filaments and has six tandem copies (S1–S6) of the gelsolin core domain. Gelsolin is the most potent actin filament-severing protein, and its severing activity is upregulated by Ca^{2+}. After severing, gelsolin remains attached to the plus-end of the filaments, creating short actin filaments that can not reanneal or elongate from their plus-ends and thus leading to the disassembly of actin filaments. In addition to promoting actin dynamics, gelsolin has a key role in apoptosis and its expression is aberrant in many cancer cells. Gelsolin is also detected in the blood circulation. Plasma gelsolin has been suggested to clear actin from the blood and to carry lipids (Kwiatkowski 1999; Sun et al. 1999).

Reducing Ca^{2+} concentrations does not simply cause the uncapping of gelsolin-capped actin filaments (McLaughlin et al. 1993). Phosphoinositides, especially PI(4,5)P$_2$, are the only known agents that can dissociate gelsolin from actin in vitro and also inhibit its severing activity (Janmey et al. 1987). Gelsolin binds PI(4,5)P$_2$ with micromolar affinity, and the binding is enhanced by Ca^{2+} and low pH (Lin et al. 1997). The binding is stereoselective and enhanced by the diacylglycerol chain: Gelsolin prefers PI(4,5)P$_2$ to PI(4)P or PI(3,4,5)P$_3$ and does not bind to InsP$_3$ (Sun et al. 1999). Gelsolin family proteins also bind lysophosphatidic acid with high affinity, and this blocks gelsolin's severing activity and uncaps actin filaments (Meerschaert et al. 1998). Gelsolin can simultaneously bind one to three PI(4,5)P$_2$ molecules within lipid vesicles. The binding is highly dependent on the physical characteristics of the bilayer (Tuominen et al. 1999). In fact, gelsolin induces changes in the packing of PI(4,5)P$_2$ within membranes (Flanagan et al. 1997).

Although the N- and C-terminal halves of gelsolin are structurally homologous, only the N-terminal half is PI(4,5)P$_2$ sensitive. Two PI(4,5)P$_2$ binding sites have been identified in this region. The first is in S1, and the second is the S1-S2 linker and the N-terminal segment of S2 (Janmey et al. 1992; Yu et al. 1992). Expressing a peptide that mimics the second site impairs fibroblast motility (Chen et al. 1996), and this site is sufficient for the PI sensitivity of gelsolin (Xian and Janmey 2002). The N-terminal half of gelsolin undergoes significant structural changes on binding to PI(4,5)P$_2$. It has been proposed that PI(4,5)P$_2$ regulates gelsolin allosterically rather than competitively blocking the actin binding site on gelsolin (Xian and Janmey 2002). It is likely that the other members of the family are regulated in a similar way.

In addition to its direct effects on actin dynamics, the gelsolin-phosphoinositide interaction can also modulate lipid signaling events. Overexpressing gelsolin and another family member, CapG, in mouse fibroblasts inhibits phospholipase Cβ and phospholipase Cγ (Sun et al. 1997). Moreover, gelsolin can be isolated as a complex with phospholipases Cγ1, Cδ, and D and with PI 3-kinase (Kwiatkowski 1999). In osteoclasts, osteopontin stimulation induces binding of c-Src to gelsolin and PI 3-kinase, and this signaling complex is required for bone resorption (Chellaiah et al. 1998). Another family member, villin, inhibits the activity of PLC-γ1 by sequestering PI(4,5)P$_2$, the substrate of PLC-γ1; however, tyrosine-phosphorylated villin associates with PLC-γ1 and activates its catalytic activity (Panebra et al. 2001).

CapG is a calcium-sensitive protein that caps the barbed ends and has three gelsolin core domains. Unlike gelsolin, CapG does not sever actin filaments (Southwick and DiNubile 1986). CapG binds PI(4,5)P$_2$ with a high affinity, and its overexpression has been associated with increased receptor-mediated phosphoinositide turnover and Ca^{2+} signaling (Sun et al. 1995).

3.2
Actin Monomer-Binding Proteins

Proteins that bind actin monomers regulate the size, localization, and dynamics of the actin monomer pool in cells. Three classes of actin monomer-binding proteins, profilins, ADF/cofilins, and twinfilins, are conserved throughout evolution and exist in organisms as distinct as budding yeast and humans. ADF/cofilins also interact with actin filaments and promote filament depolymerization. These proteins interact with phosphoinositides in vitro, and PI(4,5)P$_2$ inhibits the binding of these proteins to actin monomers. The ADF/cofilins are also inhibited from binding and depolymerizing filaments in vitro by phosphoinositides.

3.2.1
Profilin

Profilin is a small actin monomer-binding protein found in all eukaryotic cells investigated so far. It directly interacts with actin, phospholipids, and several proteins that contain polyproline stretches. Profilins have higher affinities for ATP-actin monomers than they do for ADP-actin monomers, and they enhance the nucleotide exchange on actin monomers. In the absence of filament ends profilin acts as a monomer-sequestering protein, but in the presence of free plus-ends it promotes actin polymerization (Pantaloni and Carlier 1993; Didry et al. 1998; Pollard et al. 2000).

PI(4,5)P$_2$ disrupts the actin-profilin complex in vitro (Lassing and Lindberg 1985). The affinities of PI(4,5)P$_2$ for profilin are 1–500 µM, depending on the species and profilin isoform (Schluter et al. 1997). Profilin also interacts with PI(3,4)P$_2$ and PI(3,4,5)P$_3$ and appears to have higher affinities for PI(3,4,5)P$_3$ than PI(4,5)P$_2$. Moreover, PI(3,4,5)P$_3$ more efficiently inhibits profilin's actin monomer-sequestering-activity than PI(4,5)P$_2$ (Lu et al. 1996).

Basic amino acids usually mediate the interactions of actin-binding proteins to the negatively charged headgroup of the phosphoinositide. In agreement with this, the more positively charged *Acanthamoeba* profilin II has a higher affinity for PI(4,5)P$_2$ than the more acidic profilin I (Machesky et al. 1990). The more basic human profilin I also interacts better with PI(4,5)P$_2$ than profilin IIa (Lambrechts et al. 1997, 2000). The identity of the amino acids responsible for profilin's binding to PI(4,5)P$_2$ is still a matter of debate, because there are some discrepancies between studies on vertebrate profilins and profilins from lower eukaryotes. In human profilin I, two distinct regions are vital for interacting with PI(4,5)P$_2$ (Lambrechts et al. 2002; Skare and Karlsson 2002). These sites are approximately 31 Å apart (Lambrechts et al. 2002): One overlaps with the actin-binding surface, and the other is part of the poly-l-proline binding helix (Lambrechts et al. 2002; Skare and Karlsson 2002) (Fig. 3A). This explains the observed competitions between actin and PI(4,5)P$_2$ (Lassing and Lindberg 1985) and between poly-l-proline and PI(4,5)P$_2$ for binding to profilin (Lambrechts et al. 1997). Allosteric changes may contribute to these competitive interactions, because the interaction of profilin with PI(4,5)P$_2$ or poly-l-proline induces conformational changes in profilin (Lu et al. 1996; Lambrechts et al. 1997). Specifically, the binding of PI(4,5)P$_2$ increases the α-helical content of profilin (Lu et al. 1996; Raghunathan et al. 1992). It is not known whether the invertebrate profilins associate with PI(4,5)P$_2$ in a similar manner, because they lack an analogous basic amino acid of the second PI(4,5)P$_2$ interaction site of human profilin (Lambrechts et al. 2002).

Profilin protects PI(4,5)P$_2$ from hydrolysis by phospholipase Cγ unless the phospholipase is activated (Goldschmidt-Clermont et al. 1990, 1991). PI(3,4)P$_2$ and PI(3,4,5)P$_3$ are protected from hydrolysis by profilin even if the phospholipase is activated, perhaps because they have higher affinities for profilin than PI(4,5)P$_2$ (Lu et al. 1996). Profilin directly interacts with PI 3-kinase (Bhargavi et al. 1998) and upregulates its activity (Singh et al. 1996). Some of the effects of profilin on the phosphoinositide cycle might be indirect, because profilin also interacts with several members of various signal transduction pathways through its poly-l-proline binding site (Schluter et al. 1997).

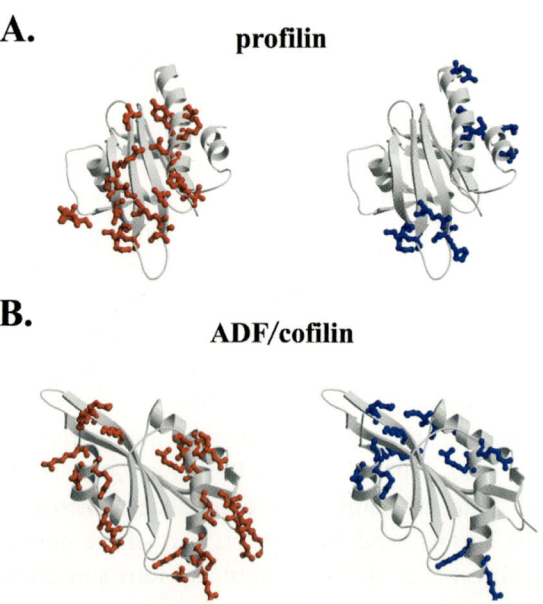

Fig. 3A, B Comparison of the actin- and PI(4,5)P$_2$-binding sites of profilin and cofilin. Ribbon diagrams of bovine profilin (**A**) and yeast cofilin (**B**). The side chains of residues shown to be important for actin and PI(4,5)P$_2$ binding are indicated by *red* and *blue*, respectively. It is important to note that the actin- and PI(4,5)P$_2$–binding sites overlap on the surfaces of cofilin and profilin, explaining why the actin-related activities of these proteins can be inhibited by PI(4,5)P$_2$. [The figures were generated with the programs MOLSCRIPT (Kraulis 1991) and RASTER3D (Merrit et al. 1994)]

3.2.2
ADF/Cofilin

ADF/cofilins are small proteins composed of a single ADF-homology domain. They are ubiquitous in eukaryotes and interact with actin monomers and filaments, preferring ADP-actin over ATP-actin (Maciver and Weeds 1994; Carlier et al. 1997; Blanchoin and Pollard 1998, 1999). In cells, ADF/cofilins enhance filament turnover by depolymerizing filaments from their minus-ends (Carlier et al. 1997) and by severing actin filaments and thus creating new filament ends (Blanchoin and Pollard 1999). ADF/cofilins are essential for the cellular processes that require a dynamic actin cytoskeleton: motility, cytokinesis, phagocytosis, and fluid-phase endocytosis (Bamburg 1999).

Some ADF/cofilins are regulated by pH and phosphorylation (Bamburg 1999). The only common regulatory mechanism for all ADF/cofilins appears to be phosphoinositide binding, which inhibits the actin-binding and depolymerization activities of ADF/cofilins (Yonezawa et al. 1990). ADF/cofilins interact with several phosphoinositides with different affinities: Yeast cofilin's affinities for phospholipids are $PI(4,5)P_2$ $=PI(3,4,5)P_3>PI(3,4)P_2>PI(4)P$. The polar headgroup, IP_3, of $PI(4,5)P_2$ does not interact with yeast cofilin (Ojala et al. 2001) or with ADF3 of *Zea mays* (Gungabissoon et al. 1998). This suggests that, in addition to the polar headgroup, the fatty acid chains of phosphoinositides also contribute to the interaction with ADF/cofilin.

The most highly conserved region in ADF/cofilins is the long α-helix, and the basic residues in this region are important for ADF/cofilin interactions with actin monomers, filaments, and $PI(4,5)P_2$ (Moriyama et al. 1992; Lappalainen et al. 1997; Van Troys et al. 2000). However, mutagenizing yeast cofilin showed that the $PI(4,5)P_2$-binding interface consists of a large number of residues that are scattered around the positively charged face of cofilin (Ojala et al. 2001). These residues are located quite far away from each other in the three-dimensional structure, indicating that cofilin might interact with more than one $PI(4,5)P_2$ molecule or that the interactions with $PI(4,5)P_2$ are dynamic and do not take place at just one specific site on cofilin. The $PI(4,5)P_2$ binding sites on yeast cofilin overlap with the areas that are important for actin binding, explaining why $PI(4,5)P_2$ inhibits the actin-related activities of ADF/cofilins (Fig. 3B) (Ojala et al. 2001).

The biological relevance of ADF/cofilins interacting with $PI(4,5)P_2$ is controversial, because other mechanisms such as phosphorylation also regulate ADF/cofilin activities. Microinjecting a nonphosphorylatable cofilin mutant into cultured myotubes caused the formation of cofilin-actin rods; this rod formation was prevented when the cofilin was treated with $PI(4,5)P_2$ before injection, indicating that $PI(4,5)P_2$ can regulate cofilin activity in vivo (Bamburg 1999). Expressing a mutant cofilin that is defective in $PI(4,5)P_2$ binding results in the appearance of thick actin bars in yeast cells; however, this mutant also has small defects in interactions with actin monomers and with Aip1, an actin-binding protein (Rodal et al. 1999; Ojala et al. 2001). Therefore, it is difficult to judge whether the phenotype resulted from specific defects in $PI(4,5)P_2$ binding.

3.2.3
Twinfilin

Twinfilin is a ubiquitous protein that binds actin monomers and is composed of two ADF-homology domains. It forms a 1:1 complex with actin monomers, and in vitro it prevents nucleotide exchange and actin filament assembly. Unlike ADF/cofilin, twinfilin does not interact with actin filaments. Deletion and overexpression phenotypes in yeast, *Drosophila*, and mammalian cells indicate that twinfilin is a universal regulator of actin filament dynamics in eukaryotic cells (Goode et al. 1998; Vartiainen et al. 2000; Wahlström et al. 2001). It has been proposed that twinfilin functions to localize actin monomers, in their inactive ADP-form, to the sites of rapid actin assembly in cells (Palmgren et al. 2002).

Twinfilin interacts with various phosphoinositides, such as PI(4)P, PI(3,4)P_2, PI(4,5)P_2, and PI(3,4,5)P_3, in native gel electrophoresis. The interaction is strongest with PI(4,5)P_2 and PI(3,4,5)P_3. PI(4,5)P_2 prevents twinfilin from binding actin monomers, and this may be a mechanism to downregulate twinfilin's sequestration of actin monomers in regions of rapid filament nucleation and assembly (Palmgren et al. 2001). Twinfilin does not interact with IP$_3$, like ADF/cofilin, suggesting that the fatty acid chains of PI(4,5)P_2 are also important for interactions with twinfilin (Palmgren et al. 2001). ADF/cofilin and twinfilin both have an ADF-homology domain, so it is possible that they interact with phosphoinositides through similar interfaces.

3.3
Proteins That Promote Actin Filament Nucleation

3.3.1
WASP/N-WASP

Arp2/3 is a nucleator of actin filaments and plays a key role in the initiation of actin polymerization. The main regulators of Arp2/3 activityare proteins of the WASP family. There are five mammalian WASP family members: WASP, N-WASP, Scar 1, Scar 2, and Scar 3. Both N-WASP and WASP are crucial in linking the signals from PI(4,5)P_2 and Cdc42 to the actin cytoskeleton, whereas the Scar proteins are effectors of the small GTPase Rac (Thrasher 2002). Because it is not known whether Scar proteins interact with phosphoinositides, these proteins are not discussed further here.

In unstimulated cells, WASP and N-WASP exist in inactive autoinhibited states, in which the N- and C-termini of a protein interact (Miki and Takenawa 1998; Kim et al. 2000). Binding of regulatory proteins to the N-termini relieves this inhibition and activates Arp2/3 and actin filament nucleation. Cdc42 enhances the ability of full-length recombinant N-WASP to activate the Arp2/3 complex in vitro, and this effect is further enhanced by PI(4,5)P_2. Cdc42 and PI(4,5)P_2 reduce the affinity between the N- and C-termini of the molecule (Rohatgi et al. 2000; Prehoda et al. 2000). On the other hand, purified native WASP is stimulated by PI(4,5)P_2 alone but not by Cdc42. However, Cdc42, when both prenylated and GTP-bound, augments PI(4,5)P_2 stimulation. Therefore, the synergy between PI(4,5)P_2 and Cdc42 requires both to be at a membrane surface (Higgs and Pollard 2000).

N-WASP can be activated independently of Cdc42. Nck is an adaptor protein that recruits proteins to sites of tyrosine phosphorylation. Binding of Nck to the proline-rich region of N-WASP stimulates the nucleation activity of Arp2/3 in vitro. This activity is further stimulated by PI(4,5)P_2 (Rohatgi et al. 2001), suggesting that PI(4,5)P_2 is a common mechanism for recruiting WASP and N-WASP to the membrane, where they can then be activated by several signaling pathways (Fig. 4B).

The EVH1/WH1-domain is the N-terminal domain of WASP and N-WASP, and it shows similarity to the phospholipid-binding PH domain (Fedorov et al. 1999; Prehoda et al. 2000), but biochemical evidence indicates that this domain does not bind PI(4,5)P_2. The binding site for PI(4,5)P_2 in WASP and N-WASP is still unresolved; however, N-WASP's lysine-rich basic region close to the GTPase binding site appears to be important for PI(4,5)P_2 interactions (Higgs and Pollard 2000; Rohatgi et al. 2000).

3.3.2
Cortactin

Cortactin binds actin filaments and activates filament nucleation through the Arp2/3 complex (Weed and Parsons 2001). It has also been reported to cross-link actin filaments (He et al. 1998). Cortactin is a substrate of many nonreceptor tyrosine kinases, and it interacts with several proteins via its SH3 domain. Cortactin is thought to be a scaffold protein

A.

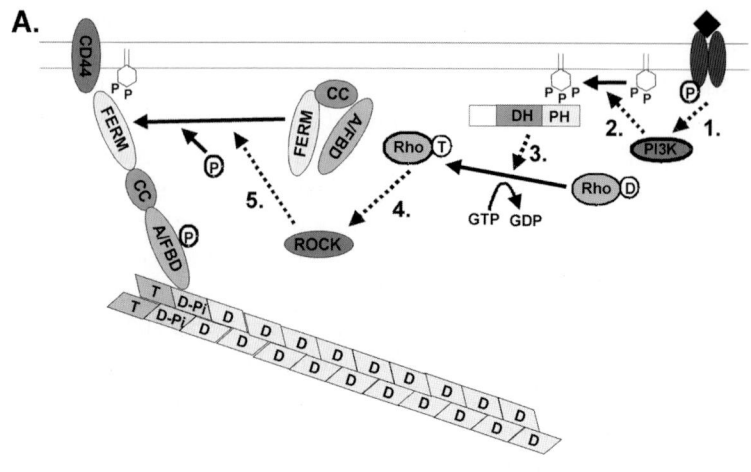

B.

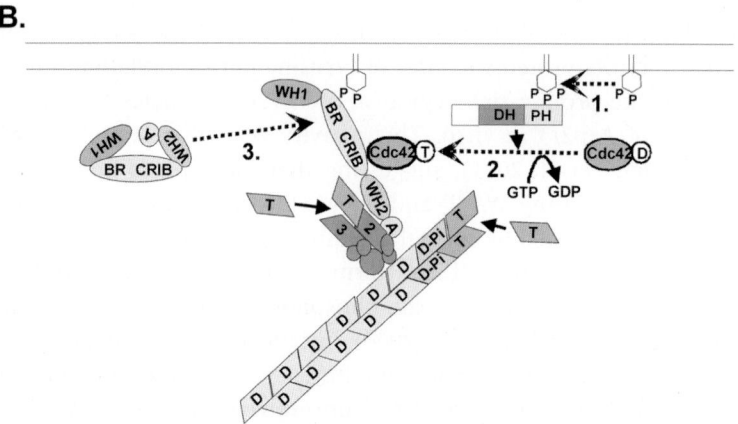

Fig. 4A, B Activation of ERM and WASP family proteins by PI(4,5)P$_2$. **A** ERM proteins are composed of a FERM (four-point one, ezrin, radixin, moesin), CC (coiled-coil), and A/FBD (actin and FERM domain binding) domains. In the absence of PI(4,5)P$_2$ ERM proteins are in an inactive from, in which the C-terminal actin-binding site of the protein is associated with the FERM domain. Phosphorylation of ERM protein at the A/FBD domain and interaction with PI(4,5)P$_2$ promote a conformational change that leads to the activation of the protein. The activated ERM proteins serve as a link between certain membrane proteins (such as CD44) and the actin cytoskeleton. Several kinases, including ROCK, phosphorylate ERM proteins. The activity of ROCK kinase is regulated by Rho. Rho may be activated through PI(3,4,5)P$_3$ responsive upregulation of its guanosine nucleotide exchange factor, Vav. **B** WASP family proteins are composed of WH1 (WASP homology domain 1), BR (basic region), CRIB (Cdc42/Rac-binding), WH2 (WASP homology domain 2), and A (acidic)

in various large protein complexes that regulate cell adhesion, actin assembly, and endocytosis (Weed and Parsons 2001).

PI(4,5)P$_2$ inhibits the actin cross-linking activity of cortactin. The putative PI(4,5)P$_2$ binding motifs are within the actin-binding repeats and include several basic residues. PI(4,5)P$_2$ is also involved in the dissociation of cortactin from the actomyosin II complex during Ras-induced malignant transformation (He et al. 1998).

3.4
Actin Cross-Linking Proteins

s assemble actin filaments into higher-order structures such as filament bundles and networks. Some of these proteins also facilitate the attachment of actin filaments to membranes and to cell-cell and cell-matrix contacts. Phosphoinositides have variable effects on the activities of different actin filament cross-linking proteins.

3.4.1
α-Actinin

α-Actinin is a major actin filament cross-linking protein in both muscle and nonmuscle cells. Besides actin, α-actinin also interacts with many cytoskeletal and signaling proteins. For example, α-actinin interacts with several focal adhesion and cell-cell contact proteins: vinculin, zyxin, and integrins (Young and Gautel 2000).

α-Actinin binds to membranes that contain negatively charged phospholipids but not to membranes that are composed only of neutral lipids. This association affects the secondary structure of α-actinin (Han

⬅—————————————————————————

domains. In the absence of PI(4,5)P$_2$ and activated Cdc42, WASP exists in an inactive form, where the N- and C-terminal regions of the protein interact with each other, thus masking the actin- and Arp2/3 binding sites. PI(4,5)P$_2$ and activated (GTP form) Cdc42 promote a conformational change in WASP, which leads to its activation and subsequent nucleation and assembly of new actin filaments through Arp2/3 complex. The acidic domain of the activated WASP interacts with Arp2/3, whereas the WH2 domain promotes actin assembly by interacting with actin monomers. The activity of the small GTPase Cdc42 can be upregulated by PI(3,4,5)P$_3$ similarly as described above for Rho

et al. 1997). PI(4,5)P$_2$ dramatically increases the cross-linking activity of α-actinin (Furuhashi et al. 1992) and controls the incorporation of α-actinin into the cytoskeleton. The actin-associated fraction of α-actinin contains a bound PI(4,5)P$_2$, whereas the cytosolic α-actinin is PI(4,5)P$_2$ free (Fukami et al. 1994). The PI(4,5)P$_2$ binding site on α-actinin is located in the second CH domain and is a stretch of evenly spaced basic amino acids. A homologous region is also found in spectrin and in the PH domains of several proteins (Fukami et al. 1996).

In muscles, α-actinin cross-links actin filaments into a tetragonal lattice and binds to two sarcomeric ruler proteins, titin and nebulin. Phospholipids are components of the Z-disk (Bullard et al. 1990), and PI(4)P-5-kinase colocalizes with α-actinin at the Z-disk. Phosphatidylserine and PI(4,5)P$_2$ activate the binding of α-actinin to titin by relieving a pseudoligand interaction between the subunits of the α-actinin dimer. The PI(4,5)P$_2$-bound α-actinin can simultaneously bind actin and titin, so PI(4,5)P$_2$ aids the targeting of α-actinin into the Z-disk (Young and Gautel 2000). Interestingly, PI(4,5)P$_2$ inhibits a fragment of titin from binding actin (Astier et al. 1998).

PI 3-kinase binds to α-actinin through its p85 subunit, and this interaction may be upregulated by PI(4,5)P$_2$ (Shibasaki et al. 1994). The product of PI 3-kinase, PI(3,4,5)P$_3$, binds to α-actinin and disrupts its interaction with the integrin β subunit. This leads to the redistribution of α-actinin from focal adhesions and actin stress fibers to the cytoplasm. This may be a mechanism to restructure focal adhesions by signaling through PI 3-kinase and PI(3,4,5)P$_3$ (Greenwood et al. 2000).

3.4.2
Filamin

Filamins are a family of high-molecular-weight cytoskeletal proteins that organize actin into three-dimensional networks. Filamins also interact with several transmembrane proteins (e.g., integrins), with signaling proteins, and with other cytoskeletal proteins. Filamin may participate in the anchoring of cortical actin and stress fibers to transmembrane receptors and could function as a scaffold for signaling complexes (van der Flier and Sonnenberg 2001; Stossel et al. 2001).

Filamin, like some other actin-binding proteins, binds to acidic phospholipids, but with little or no selectivity for inositol lipids (Tempel et al. 1994). However, in contrast to α-actinin, the cross-linking activity of filamin is inhibited by PI(4,5)P$_2$ (Furuhashi et al. 1992). The mechanism of this inhibition is not known, but because filamin can insert into phospholipid monolayers, PI(4,5)P$_2$ may sequester filamin away from the sites of actin filament cross-linking (Tempel et al. 1994).

3.4.3
Cortexillin

Cortexillins are actin-bundling proteins that organize actin filaments into antiparallel bundles and three-dimensional networks. In *Dictyostelium*, cortexillin plays a role in cytokinesis and participates in forming the cleavage furrow during mitotic division (Faix et al. 1996; Weber et al. 1999). Cortexillins also contribute to the mechanical properties of the cell cortex (Simson et al. 1998).

Cortexillin has an actin-binding domain (the CH domain of α-actinins and spectrin), but the actin-bundling activity is in a separate C-terminal domain. This region contains a lysine-rich nonapeptide sequence that mediates the binding of cortexillin to PI(4,5)P$_2$. These nine amino acids also enhance the affinity of cortexillin for targets that reside in the cell cortex. PI(4,5)P$_2$ inhibits the actin-bundling activity of cortexillin, although the PI(4,5)P$_2$-cortexillin complex appears to still interact with actin filaments (Stock et al. 1999).

3.5
Proteins Linking the Actin Cytoskeleton to the Plasma Membrane

Proteins linking the actin cytoskeleton to the plasma membrane are often controlled by phosphoinositides. Generally the binding to phosphoinositide [usually PI(4,5)P$_2$] induces a conformational change in linking protein, thereby exposing the binding sites for actin or other binding proteins.

3.5.1
ERM Proteins

The ERM family and the related tumor suppressor protein merlin are part of the protein 4.1 superfamily. These proteins link actin filaments at the cell cortex to membrane proteins. The regulated attachment of membrane proteins to actin filaments is essential for many fundamental processes, including the determination of cell shape and surface structures, cell adhesion, motility, cytokinesis, phagocytosis, and the integration of membrane transport with signaling pathways. One of the most studied processes in which ERM proteins are involved is microvilli formation and breakdown (Bretscher et al. 2002).

ERM proteins are composed of an N-terminal FERM (four-point one, ezrin, radixin, moesin) domain, a putative coiled-coil region, and a C-terminal A/FBD (ERM actin- and FERM binding) domain. In the inactive state, the N-terminal FERM and C-terminal A/FBD domains interact and thus mask some important interaction-sites of the protein, such as the actin-binding site. Activation of ERM proteins requires these two domains to separate. Separation is regulated by threonine phosphorylation, but interaction with $PI(4,5)P_2$ is also necessary for activation. Binding to $PI(4,5)P_2$ is mediated by the N-terminal FERM domain of these proteins (Bretscher et al. 2002). According to crystal structures, the FERM domains of radixin and moesin have three subdomains. Domain A consists of a long α-helix and a five-stranded mixed β-sheet similar to the ubiquitin fold. Domain B is composed of α-helical structures, and domain C has a fold similar to the PH domains (Pearson et al. 2000; Hamada et al. 2000). According to the crystal structure of the radixin FERM domain complexed to IP_3, the amino acid residues that directly contact IP_3 are in a basic cleft between subdomains A and C (Hamada et al. 2000). However, mutational studies revealed that additional residues are involved in $PI(4,5)P_2$ binding (Barret et al. 2000).

The ERM family anchors to the plasma membrane in different ways. They can indirectly bind certain transmembrane proteins, such as the cystic fibrosis transmembrane conductance regulator (Short et al. 1998) or the Na^+/H^+ antiporter (Murthy et al. 1998; Yun et al. 1998), via an adaptor protein such as EBP50 (Reczek et al. 1997). ERM proteins can also directly interact with the cytoplasmic tails of membrane proteins with a single transmembranous domain such as CD43, CD44, ICAM-1, ICAM-2, and ICAM-3. These interactions are facilitated by $PI(4,5)P_2$ (Hirao et al. 1996). Biochemical studies have also shown that $PI(4,5)P_2$, or some other lipid or detergent, is required to unmask the F-actin binding site in ERM proteins (Nakamura et al. 1999). Furthermore, cells depleted of phosphoinositides by neomycin treatment lose cell-surface microvilli, presumably because $PI(4,5)P_2$ is required for ERM protein activation in vivo (Yonemura et al. 2002). The actual molecular mechanism of ERM activation is unknown, but it is hypothesized that $PI(4,5)P_2$ recruits ERM proteins to the plasma membrane, locates them close to Rho-kinase or protein kinase-Cθ, and leads to the phosphorylation and subsequent activation of these proteins (Fig. 4A) (Bretscher et al. 2002).

3.5.2
Talin

Talin is a structural component of focal adhesion sites and is involved in multiple interactions at the cytoplasmic face of cell or matrix contacts. Talin is a major link between integrins and the actin cytoskeleton and is important to focal adhesion assembly. Talin has a 50-kDa head and 205-kDa rod domains. The head domain contains a major integrin-binding site and a region homologous to the FERM domain of ERM proteins (Rees et al. 1990). The FERM domain is the phosphoinositide binding site in ERM proteins, and, accordingly, talin also interacts with PI(4,5)P$_2$. The high-affinity interaction between talin and engaged integrins is promoted by PI(4,5)P$_2$ (Martel et al. 2001). PI(4,5)P$_2$ might induce a conformational change in talin, thereby exposing the binding sites for integrins and contributing to the assembly of focal adhesions. PI(4,5)P$_2$ is also necessary for talin to remain in focal adhesions (Martel et al. 2001). Recent studies also show that talin recruits at least some types of PI(4)P 5-kinase to the focal contacts; this recruitment defines a mechanism for the spatial generation of PI(4,5)P$_2$ at focal adhesions (Di Paolo et al. 2002; Ling et al. 2002).

3.5.3
Vinculin

Vinculin is also in focal adhesions and mediates the linkage of cadherins and integrins to the actin cytoskeleton, an overall role similar to talin's. Talin appears to be important for forming focal adhesions, whereas vinculin may stabilize adhesion in focal contacts (Volberg et al. 1995).

Vinculin has an N-terminal 90-kDa globular domain connected to a 30-kDa tail domain by a short proline-rich region. The recruitment and assembly of vinculin into focal adhesions is downregulated by strong intramolecular interactions between the head and tail domains. This interaction associates the talin- and α-actinin-binding sites of the head domain to the VASP-binding site of the proline-rich linker region and to the F-actin-binding site on the tail domain. Binding of PI(4,5)P$_2$ to the tail domain disrupts the head-to-tail association and unmasks the talin-binding site (Gilmore and Burridge 1996), but $PI(4,5)P_2$-activated vinculin does not bind to actin (Steimle et al. 1999). Activation of actin binding involves changes that are distinct from or in addition to those created by $PI(4,5)P_2$ activation (Steimle et al. 1999).

The C-terminal tail domain of vinculin contains the $PI(4,5)P_2$ binding site and is a bundle of five α-helices with N- and C-terminal extensions. These five amphipathic helices are connected by short loops and adopt an antiparallel topology with their hydrophobic residues buried inside the bundle. The last helix is followed by a C-terminal arm that terminates into a five-residue hydrophobic hairpin (Niggli 2001). Intriguingly, this helical bundle resembles that of exchangeable apolipoproteins, which can reversibly unfold into an extended membrane-bound form. Similarly, the tail domain of vinculin appears to unfold in the presence of lipids: exposing hydrophobic residues that allow extensive interactions with a bilayer (Niggli 2001).

3.6
Phosphoinositide-Responsive Activators of the Rho Family GTPases

The Rho family GTPases (namely, Rho, Rac, and Cdc42) are central proteins in mediating various signals to the actin cytoskeleton. Rho activity is required for stress fiber formation (Ridley and Hall 1992), Rac promotes the formation of lamellipodia (Ridley et al. 1992), and Cdc42 promotes the formation of filopodia (Nobes and Hall 1995). Phosphoinositides, especially $PI(3,4,5)P_3$, are important regulators of the activity of Rho family GTPases.

Each Rho GTPase has a unique set of effector proteins that mediate their effects to the actin cytoskeleton. Over 30 effector proteins have been identified. Rho-associated kinase (ROCK), mDia, and LIM kinase are Rho effectors. On activation ROCK induces the phosphorylation of myosin light chains (MLC), which stimulates the actin-activated ATPase activity of myosin II and promotes the assembly of actomyosin filaments (Hirose et al. 1998). The ERM proteins are also ROCK substrates (Bishop and Hall 2000). ROCK alone is not able to stimulate the formation of stress fibers but requires the assistance of mDia. mDia is a member of the formin-homology (FH) family of proteins, and in combination with profilin promotes actin polymerization (Watanabe et al. 1997). ROCK also activates the LIM kinase to phosphorylate ADF/cofilins, thus leading to a stabilization of actin filaments (Maekawa et al. 1999).

Rac and Cdc42 have several common effectors, but some are unique to either GTPase (Bishop and Hall 2000). Cdc42 is able to bind to WASP and N-WASP, which are activators of the Arp2/3 complex. The activated Arp2/3 promotes nucleation of actin filaments (Rohatgi et al. 1999). Cdc42 can also induce phosphorylation of MLC, via myotonic dystrophy kinase-related Cdc42-binding kinase (MRCK) (Leung et al. 1998). MRCKα is also able to activate LIM kinases 1 and 2 and thus induce inactivation of ADF/cofilins (Sumi et al. 2001). The Cdc42 effector PAK4, a novel PAK family member, can activate LIM kinase 1 (Dan et al. 2001). Only a few Rac-specific effectors are known. Por-1 has been implicated in Rac-induced lamellipodia formation, but the mechanism is unknown (Van Aelst et al. 1996; D'Souza-Schorey et al. 1997). Also, Rac can activate Arp2/3 through proteins of the Scar/WAVE family, but the activation mechanisms are different from those of cdc42 and WASP (Eden et al. 2002). Common targets for Rac and Cdc42 include PAK kinases 1, 2, and 3. These PAKs activate LIM kinase (Arber et al. 1998; Edwards et al. 1999), and they decrease MLC phosphorylation by inactivating MLC kinase (Sanders et al. 1999). This is antagonistic to the effects of Rho and leads to a reduced assembly of actomyosin structures.

The activity of small GTPases is determined by a cycle between inactive GDP-bound and active GTP-bound forms. This cycle is regulated by guanosine nucleotide exchange factors (GEFs) and GTPase-activating proteins (GAPs). GEFs catalyze the release of GDP and the subsequent binding of GTP, whereas GAPs stimulate the usually low intrinsic GTPase activity of small GTPases, converting them from their active to inactive form (Van Aelst and D'Souza-Schorey 1997; Hall 1998). Different members of the Rho family GTPases are affected by distinct sets of different GEFs. The Dbl family of GEFs are defined by the presence of a conserved domain of ~150 amino acids, designated the Dbl homology (DH) domain. The DH domain is required for GEF activity and is invariably coupled to the PH domain, a structural module of approximately 100–120 residues (Haslam et al. 1993; Mayer et al. 1993; Musacchio et al. 1993; Gibson et al. 1994). Several PH-domains bind certain phosphorylated phosphoinositides with high specificities (Rameh et al. 1997; Kavran et al. 1998). The invariant association between PH and DH domains in GEFs suggests that the PH domain may be critical for the function of DH domains.

It has been suggested that PH domain binding to phosphoinositide would activate the GEF activity of the DH domain by an allosteric mechanism. This seems to be true for RacGEF Sos1 (Das et al. 2000) and for the Vav family proteins, which have GEF activities for multiple Rho GTPases such as RhoA, RhoG, Rac1, and Cdc42 (Crespo et al. 1997; Han et al. 1998; Schuebel et al. 1998; Movilla and Bustelo 1999; Abe et al. 2000; Liu and Burridge 2000; Zeng et al. 2000). In the Sos1 and Vav proteins, the PH domain serves as a negative regulator of DH domain function. In vitro, this negative regulatory function is promoted by $PI(4,5)P_2$ and is antagonized by $PI(3,4,5)P_3$ (Han et al. 1998; Das et al. 2000). In contrast, the binding of the PH domain of Dbl to $PI(4,5)P_2$ or $PI(3,4,5)P_3$ appears to inhibit GEF activity (Russo et al. 2001). Alternatively, the interactions between phosphoinositides and PH domains may localize the DH domain to its cellular site of action. This is the case with the Rac GEF Tiam1; however, the membrane recruitment is mediated by its additional PH domain, which is not linked to the DH domain. Membrane targeting occurs via specific binding to $PI(3,4,5)P_3$ (Michiels et al. 1997). The Tiam1 PH domain that is adjacent to the DH domain binds $PI(3)P$ and not $PI(4,5)P_2$ or $PI(3,4,5)P_3$ (Snyder et al. 2001). This binding does not affect GEF activity, and its function remains unclear (Snyder et al. 2001).

In summary, several lines of evidence suggest that phosphoinositides, especially $PI(3,4,5)P_3$, are central in regulating the activities of the Rho family GTPases; however, there is no universal mechanism for how the interaction of these phospholipids with the DH-PH domain pair in Rho GEFs promotes their activation.

3.7
Others

3.7.1
GMC Proteins

GAP43 (growth-associated protein), MARCKS (myristoylated alanine-rich C kinase substrate), and Cap23 (cytoskeleton-associated protein) (GMC proteins) form a family of actin filament-binding proteins that are enriched in neuronal cells. These proteins do not have strong sequence similarities, but they share some functional and physical properties. The N-termini of GMC-proteins contain a myristate or palmitate group that inserts into the plasma membrane, and they all contain a conserved basic region (effector domain) that binds Ca^{2+}/CaM, phospholipids, actin, and PKC. The exact biological functions of these proteins are not known, but they seem to be involved in actin-dependent morphogenetic and motile processes (Caroni 2001; McLaughlin et al. 2002).

Binding of MARCKS to membranes requires both the hydrophobic insertion of the N-terminal myristate into the bilayer and an electrostatic interaction of acidic lipids with a conserved cluster of basic residues in the effector domain (McLaughlin et al. 2002). The effector domain of MARCKS binds PI(4,5)P$_2$ with a high affinity. However, the MARCKS effector domain does not exhibit detectable specificity between PI(3,4)P$_2$ and PI(4,5)P$_2$ (Wang et al. 2001). Moreover, high ionic strength inhibits the binding (Kim et al. 1994), and the strong binding probably involves formation of an electrically neutral complex between one peptide and three to four lipid molecules. Therefore, binding of the MARCKS effector domain to PI(4,5)P$_2$ is a nonspecific electrostatic interaction (Wang et al. 2001). Interaction with Ca^{2+}/CaM or phosphorylation by PKC releases the effector domains of GMC proteins from the membrane (Arbuzova et al. 1998; Ohmori et al. 2000).

MARCKS and PI(4,5)P$_2$ are present in cells at similar concentrations. Therefore, a significant fraction of the PI(4,5)P$_2$ in the membrane could be sequestered by MARCKS. Indeed, physiological concentrations of the intact protein (10 µM) or a peptide corresponding to the effector domain inhibit PLC activity by sequestering PI(4,5)P$_2$ away from the catalytic domain of PLC (Glaser et al. 1996; Wang et al. 2001). Furthermore, MARCKS, GAP43, and CAP23 colocalize with PI(4,5)P$_2$ in fixed cells and promote the clustering and retention of PI(4,5)P$_2$ in these sites. These functions are dependent on the effector domains of these proteins, and constructs lacking the effector domain can function as dominant-negative inhibitors of plasmalemmal PI(4,5)P$_2$ modulation (Laux et al. 2000). Overexpression of MARCKS results in an increase in cellular PI(4,5)P$_2$ levels, indicating that cells aim to maintain a constant concentration of free PI(4,5)P$_2$. Overexpression of GAP43 or CAP23 does not produce this heightened level of PI(4,5)P$_2$, suggesting that they might be less important PI(4,5)P$_2$ buffers than MARCKS. Because all GMC proteins are localized to small cholesterol-enriched rafts in cells and are able to bind PI(4,5)P$_2$, they have important roles in localizing PI(4,5)P$_2$ to these membrane rafts (Laux et al. 2000).

MARCKS, and probably also GAP43 and CAP23, may regulate the actin cytoskeleton by two mechanisms. First, these proteins directly bind to actin so they may cross-link actin filaments to the plasma membrane. Phosphorylation by PKC or binding of Ca^{2+}/CaM would then result in the release of GMC proteins along with actin from the plasma membrane. Alternatively, binding of GMC proteins to the plasma membrane may sequester $PI(4,5)P_2$ in cells. Phosphorylation by PKC or binding to Ca^{2+}/CaM could release GMC proteins from the membrane and thus allow $PI(4,5)P_2$ to interact with other proteins that modulate the actin cytoskeleton. Therefore, GMC proteins may bind $PI(4,5)P_2$ globally and release it locally, producing localized changes in the actin cytoskeleton (Arbuzova et al. 2002; McLaughlin et al. 2002).

3.7.2
Myosin X

Myosin X (Myo10) is a recently identified unconventional myosin. It is a two-headed nonprocessive motor that moves toward the plus-end of the actin filament (Homma et al. 2001). Myosin X contains three tandem PH domains in its tail domain, a feature that is not found in other known myosins. The PH domains in myosin X have an unusual organization in that the first PH domain (PH1) is split by the insertion of the second PH domain (PH2). Of the three PH domains of myosin X, the PH2 fits best to the consensus sequence for binding the PI 3-kinase product and it was shown to interact with $PI(3,4)P_2$ and/or $PI(3,4,5)P_3$ in an elegant yeast genetic interaction assay (Isakoff et al. 1998). No additional in vitro studies on the phosphoinositide-binding specificities of these PH domains have been carried out so far. However, according to sequence analysis of the myosin X PH2 domain, this domain is expected to bind $PI(3,4,5)P_3$ with high specificity (Isakoff et al. 1998; Berg et al. 2000; Ferguson et al. 2000; Thomas et al. 2001). In agreement with these data, the localization of myosin X to phagocytic cups is abolished by the PI 3-kinase inhibitor wortmannin (Cox et al. 2002).

Myosin X localizes to cell regions characterized by rapid actin polymerization, such as the edge of the lamellipodia, membrane ruffles, and filopodia tips (Berg et al. 2000). Furthermore, in macrophages myosin X is recruited to phagocytic cups, and based on mutagenesis results and studies with a PI 3-kinase inhibitor it was proposed that myosin X provides a molecular link between PI 3-kinase and pseudopod extension during phagocytosis (Cox et al. 2002). Taken together, myosin X is an example of a protein that directly binds to PI(3,4,5)P$_3$ and actin filaments. As discussed above, most other actin-associated proteins that are regulated by PI(3,4,5)P$_3$ display their effects on the actin cytoskeleton indirectly through, for example, Rho family GTPases.

4
The Molecular Mechanism of Phosphoinositide-Promoted Actin Assembly

A large number of studies have demonstrated that PI(4,5)P$_2$ and PI(3,4,5)P$_3$ promote actin filament assembly. Overexpression of PI(4)P 5-kinase in various mammalian cell types results in an accumulation of actin filaments, whereas downregulating PI(4,5)P$_2$ decreases actin polymerization in cells (see, e.g., Shibasaki et al. 1997; Sakisaka et al. 1997; Rozelle et al. 2000; Raucher et al. 2000; Yamamoto et al. 2001). Similarly, increases in cellular PI(3,4,5)P$_3$ levels promote actin polymerization and cell motility, whereas inhibition of PI(3,4,5)P$_3$ production decreases cell motility (Keely et al. 1997; Shaw et al. 1997; Derman et al. 1997; Niggli 2000). In agreement with the role of phosphoinositides in promoting actin assembly, the activities of proteins that increase actin assembly, such as the WASPs, are upregulated by PI(4,5)P$_2$. In contrast, the activities of proteins that inhibit actin assembly or promote filament disassembly, such as capping protein, gelsolin, and ADF/cofilin and twinfilin, are downregulated by PI(4,5)P$_2$. Similarly, PI(3,4,5)P$_3$ is also expected to promote actin assembly, because it upregulates Rho family GTPases through specific binding and activation of Rho GEFs.

In resting cells or cell regions, approximately 1% of all phospholipids at the plasma membrane is PI(4,5)P$_2$, whereas the PI(3,4,5)P$_3$ concentration is believed to be very low (Auger et al. 1989; McLaughlin et al. 2002). PI(4,5)P$_2$ in these cells is either uniformly located at the plasma membrane and some other membranes, or concentrated at PI(4,5)P$_2$-rich membrane microdomains. The actin polymerization in these cells is restricted to the proximity of the plasma membrane, because PI(4,5)P$_2$ promotes actin polymerization and also inhibits the proteins that induce disassembly. However, because of the absence of PI(3,4,5)P$_3$, the Rho family GTPases are inactive and the subsequent nucleation of new actin filaments by WASP-activated Arp2/3 complex is slow. Because Rho family GTPases also promote inactivation of ADF/cofilin through its phos-

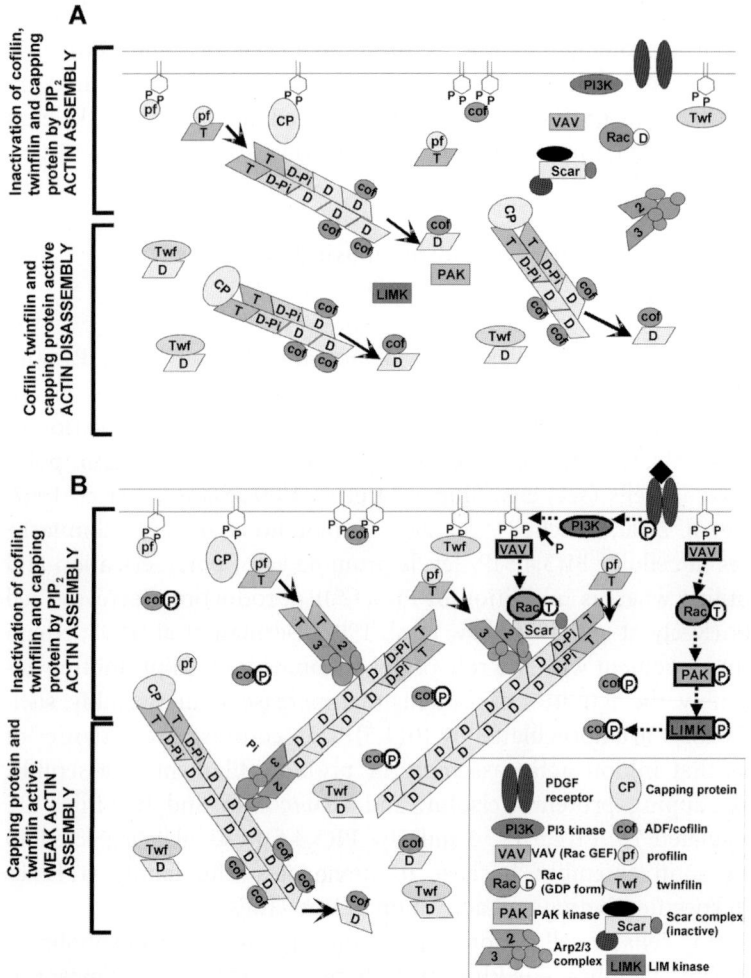

Fig. 5A, B A model for the roles of PI(4,5)P$_2$ and PI(3,4,5)P$_3$ in regulating actin filament assembly in cells. **A** PI(4,5)P$_2$ inactivates proteins such as ADF/cofilin, twinfilin, and capping protein close to the plasma membrane. Because these proteins inhibit actin assembly and promote actin filament disassembly, this leads to an increased actin filament assembly close to the PI(4,5)P$_2$-rich regions of the plasma membrane. However, in the absence of activated Rho family GTPases the nucleation of new actin filaments is slow. Furthermore, the formation of stable actin filaments in the cytoplasm is prevented by the lack of ADF/cofilin phosphorylation. **B** Certain growth factor stimuli (e.g., PDGF) activate PI 3-kinases, which leads to local production of PI(3,4,5)P$_3$. This stimulus-responsive phospholipid can activate several GEFs for Rho family GTPases, such as Vav. This leads to a formation of active Rac, which promotes actin filament nucleation through activation of Scar protein. Furthermore, Rac promotes ADF/cofilin phosphorylation through PAK and LIM kinases, thus leading to a decrease in the actin filament depolymerization rates

phorylation by LIM kinase, in the absence of PI(3,4,5)P$_3$, ADF/cofilin is active in regions other than PI(4,5)P$_2$-rich plasma membrane. This leads to a rapid depolymerization of cytoplasmic actin filaments (Fig. 5A).

After certain extracellular or intracellular stimuli, some PI 3-kinases become activated. This leads to a phosphorylation of PI(4,5)P$_2$ at position 3 and production of PI(3,4,5)P$_3$. This phospholipid directly binds to the PH domains of certain Rho GEFs and activates the guanosine nucleotide exchange activity of these GEFs, which leads to activation of a specific set of Rho family proteins. These GTPases then activate actin assembly through WASP family proteins and downregulate actin filament depolymerization through inactivation of ADF/cofilins. Thus PI(3,4,5)P$_3$, together with PI(4,5)P$_2$, promotes actin assembly and stabilization (Fig. 5B). Furthermore, the Rho family proteins also regulate actin filament cross-linking, contractility and the plasma membrane-cytoskeleton interactions and thus assemble actin filaments into specialized structures such as stress fibers.

The phosphoinositides PI(4,5)P$_2$ and PI(3,4,5)P$_3$ control the activity of various actin-associated proteins and thus promote actin filament assembly at the desired cell regions. It is also important to note that PI(4,5)P$_2$ may also control the direction of actin filaments and ensure that the fast-growing plus-ends are oriented toward the plasma membrane in cells. In the future, it will be important to elucidate, at the molecular level, how phosphoinositides promote their activities on various actin-associated proteins. Furthermore, although a large number of actin-binding proteins have been shown to interact with phospholipids in vitro, there is only a limited amount of evidence on the in vivo relevance of these interactions. Therefore, it will be important to examine the cellular roles of these protein-phospholipid interactions by genetic, cell biology, and microscopic methods.

Acknowledgments. We thank Dr. Helen Yin for providing Fig. 2 and Ville Paavilainen for Fig. 3. This work was supported by grants from Sigrid Juselius Foundation, Biocentrum Helsinki, and Academy of Finland.

References

Abe K, Rossman KL, Liu B, Ritola KD, Chiang D, Campbell SL, Burridge K, Der CJ (2000) Vav2 is an activator of Cdc42, Rac1, and RhoA. J Biol Chem 275:10141–10149

Amatruda JF, Cannon JF, Tatchell K, Hug C, Cooper JA (1990) Disruption of the actin cytoskeleton in yeast capping protein mutants. Nature 344:352-354

Arber S, Barbayannis FA, Hanser H, Schneider C, Stanyon CA, Bernard O, Caroni P (1998) Regulation of actin dynamics through phosphorylation of cofilin by LIM-kinase. Nature 393:805-809

Arbuzova A, Murray D, McLaughlin S (1998) MARCKS, membranes, and calmodulin: kinetics of their interaction. Biochim Biophys Acta 1376:369-379

Arbuzova A, Schmitz AA, Vergeres G (2002) Cross-talk unfolded: MARCKS proteins. Biochem J 362:1-12

Astier C, Raynaud F, Lebart MC, Roustan C, Benyamin Y (1998) Binding of a native titin fragment to actin is regulated by PIP2. FEBS Lett 429:95-98

Auger KR, Cantley LC (1991) Novel polyphosphoinositides in cell growth and activation. Cancer Cells 3:263-270

Auger KR, Serunian LA, Soltoff SP, Libby P, Cantley LC (1989) PDGF-dependent tyrosine phosphorylation stimulates production of novel polyphosphoinositides in intact cells. Cell 57:167-175

Bamburg JR (1999) Proteins of the ADF/cofilin family: essential regulators of actin dynamics. Annu Rev Cell Dev Biol 15:185-230

Barkalow K, Witke W, Kwiatkowski DJ, Hartwig JH (1996) Coordinated regulation of platelet actin filament barbed ends by gelsolin and capping protein. J Cell Biol 134:389-399

Barret C, Roy C, Montcourrier P, Mangeat P, Niggli V (2000) Mutagenesis of the phosphatidylinositol 4,5-bisphosphate (PIP$_2$) binding site in the NH$_2$-terminal domain of ezrin correlates with its altered cellular distribution. J Cell Biol 151:1067-1080

Berg JS, Derfler BH, Pennisi CM, Corey DP, Cheney RE (2000) Myosin-X, a novel myosin with pleckstrin homology domains, associates with regions of dynamic actin. J Cell Sci 3439-3451

Bhargavi V, Chari VB, Singh SS (1998) Phosphatidylinositol 3-kinase binds to profilin through the p85 alpha subunit and regulates cytoskeletal assembly. Biochem Mol Biol Int 46:241-248

Bishop AL, Hall A (2000) Rho GTPases and their effector proteins. Biochem J 348:241-255

Blanchoin L, Pollard TD (1998) Interaction of actin monomers with *Acanthamoeba* actophorin (ADF/cofilin) and profilin. J Biol Chem 273:25106-25111

Blanchoin L, Pollard TD (1999) Mechanism of interaction of *Acanthamoeba* actophorin (ADF/cofilin) with actin filaments. J Biol Chem 274:15538-15546

Bresnick AR (1999) Molecular mechanisms of nonmuscle myosin-II regulation. Curr Opin Cell Biol 11:26-33

Bretscher A, Edwards K, Fehon RG (2002) ERM proteins and merlin: integrators at the cell cortex. Nat Rev Mol Cell Biol 3:586-599

Bullard B, Sainsbury G, Miller N (1990) Digestion of proteins associated with the Z-disc by calpain. J Muscle Res Cell Motil 11:271-279

Carlier MF, Laurent V, Santolini J, Melki R, Didry D, Xia GX, Hong Y, Chua NH, Pantaloni D (1997) Actin depolymerizing factor (ADF/cofilin) enhances the rate of filament turnover: implication in actin-based motility. J Cell Biol 136:1307-1322

Carlier MF, Ressad F, Pantaloni D (1999) Control of actin dynamics in cell motility. Role of ADF/cofilin. J Biol Chem 274:33827–33830

Caroni P (2001) New EMBO members' review: actin cytoskeleton regulation through modulation of PI(4,5)P$_2$ rafts. EMBO J 20:4332–4336

Chellaiah M, Fitzgerald C, Alvarez U, Hruska K (1998) c-Src is required for stimulation of gelsolin-associated phosphatidylinositol 3-kinase. J Biol Chem 273:11908–11916

Chen P, Murphy-Ullrich JE, Wells A (1996) A role for gelsolin in actuating epidermal growth factor receptor-mediated cell motility. J Cell Biol 134:689–698

Chong LD, Traynor-Kaplan A, Bokoch GM, Schwartz MA (1994) The small GTP-binding protein Rho regulates a phosphatidylinositol 4-phosphate 5-kinase in mammalian cells. Cell 79:507–513

Chung CY, Funamoto S, Firtel RA (2001) Signaling pathways controlling cell polarity and chemotaxis. Trends Biochem Sci 26:557–566

Cooper JA, Schafer DA (2000) Control of actin assembly and disassembly at filament ends. Curr Opin Cell Biol 12:97–103

Cox D, Berg JS, Cammer M, Chinegwundoh JO, Dale BM, Cheney RE, Greenberg S (2002) Myosin X is a downstream effector of PI(3)K during phagocytosis. Nat Cell Biol 4:469–477

Crespo P, Schuebel KE, Ostrom AA, Gutkind JS, Bustelo XR (1997) Phosphotyrosine-dependent activation of Rac-1 GDP/GTP exchange by the vav proto-oncogene product. Nature 385:169–172

Czech MP (2000) PIP2 and PIP3: complex roles at the cell surface. Cell 100:603–606

Damen JE, Liu L, Rosten P, Humphries RK, Jefferson AB, Majerus PW, Krystal G (1996) The 145-kDa protein induced to associate with Shc by multiple cytokines is an inositol tetraphosphate and phosphatidylinositol 3,4,5-triphosphate 5-phosphatase. Proc Natl Acad Sci USA 93:1689–1693

Dan C, Kelly A, Bernard O, Minden A (2001) Cytoskeletal changes regulated by the PAK4 serine/threonine kinase are mediated by LIM kinase 1 and cofilin. J Biol Chem 276:32115–32121

Das B, Shu X, Day GJ, Han J, Krishna UM, Falck JR, Broek D (2000) Control of intramolecular interactions between the pleckstrin homology and Dbl homology domains of Vav and Sos1 regulates Rac binding. J Biol Chem 275:15074–15081

Derman MP, Toker A, Hartwig JH, Spokes K, Falck JR, Chen CS, Cantley LC, Cantley LG (1997) The lipid products of phosphoinositide 3-kinase increase cell motility through protein kinase C. J Biol Chem 272:6465–6470

Di Paolo G, Pellegrini L, Letinic K, Cestra G., Zoncu R, Voronov S, Chang S, Guo J, Wenk MR, De Camilli P (2002) Recruitment and regulation of phosphatidylinositol phosphate kinase type 1gamma by the FERM domain of talin. Nature 420:85–89

Didry D, Carlier MF, Pantaloni D (1998) Synergy between actin depolymerizing factor/cofilin and profilin in increasing actin filament turnover. J Biol Chem 273:25602–25611

Dowler S, Currie RA, Campbell DG, Deak M, Kular G, Downes CP, Alessi DR (2000) Identification of pleckstrin-homology-domain-containing proteins with novel phosphoinositide-binding specificities. Biochem J 351:19–31

D'Souza-Schorey C, Boshans RL, McDonough M, Stahl PD, Van Aelst L (1997) A role for POR1, a Rac1-interacting protein, in ARF6-mediated cytoskeletal rearrangements. EMBO J 16:5445–5454

Eden S, Rohatgi R, Podtelejnikov AV, Mann M, Kirschner MW (2002) Mechanism of regulation of WAVE1-induced actin nucleation by Rac1 and Nck. Nature 418:790–793

Edwards DC, Sanders LC, Bokoch GM, Gill GN (1999) Activation of LIM-kinase by Pak1 couples Rac/Cdc42 GTPase signalling to actin cytoskeletal dynamics. Nature Cell Biol 1:253–259

Faix J, Steinmetz M, Boves H, Kammerer RA, Lottspeich F, Mintert U, Murphy J, Stock A, Aebi U, Gerisch G (1996) Cortexillins, major determinants of cell shape and size, are actin-bundling proteins with a parallel coiled-coil tail. Cell 86:631–642

Fedorov AA, Fedorov E, Gertler F, Almo SC (1999) Structure of EVH1, a novel proline-rich ligand-binding module involved in cytoskeletal dynamics and neural function. Nat Struct Biol 6:661–665

Ferguson KM, Kavran JM, Sankaran VG, Fournier E, Isakoff SJ, Skolnik EY, Lemmon MA (2000) Structural basis for discrimination of 3-phosphoinositides by pleckstrin homology domains. Mol Cell 6:373–384

Flanagan LA, Cunningham CC, Chen J, Prestwich GD, Kosik KS, Janmey PA (1997) The structure of divalent cation-induced aggregates of PIP2 and their alteration by gelsolin and tau. Biophys J 73:1440–1447

Franke TF, Kaplan DR, Cantley LC, Toker A (1997) Direct regulation of the Akt proto-oncogene product by phosphatidylinositol-3,4-bisphosphate. Science 275:665–668

Fruman DA, Meyers RE, Cantley LC (1998) Phosphoinositide kinases. Annu Rev Biochem 67:481–507

Fukami K, Endo T, Imamura M, Takenawa T (1994) alpha-Actinin and vinculin are PIP2-binding proteins involved in signaling by tyrosine kinase. J Biol Chem 269:1518–1522

Fukami K, Sawada N, Endo T, Takenawa T (1996) Identification of a phosphatidylinositol 4,5-bisphosphate-binding site in chicken skeletal muscle alpha-actinin. J Biol Chem 271:2646–2650

Furuhashi K, Inagaki M, Hatano S, Fukami K, Takenawa T (1992) Inositol phospholipid-induced suppression of F-actin-gelating activity of smooth muscle filamin. Biochem Biophys Res Commun 184:1261–1265

Gibson TJ, Hyvonen M, Musacchio A, Saraste M, Birney E (1994) PH domain: the first anniversary. Trends Biochem Sci 19:349–353

Gilmore AP, Burridge K (1996) Regulation of vinculin binding to talin and actin by phosphatidyl-inositol-4-5-bisphosphate. Nature 381:531–535

Glaser M, Wanaski S, Buser CA, Boguslavsky V, Rashidzada W, Morris A, Rebecchi M, Scarlata SF, Runnels LW, Prestwich GD, Chen J, Aderem A, Ahn J, McLaughlin S (1996) Myristoylated alanine-rich C kinase substrate (MARCKS) produces reversible inhibition of phospholipase C by sequestering phosphatidylinositol 4,5-bisphosphate in lateral domains. J Biol Chem 271:26187–26193

Glogauer M, Hartwig J, Stossel T (2000) Two pathways through Cdc42 couple the N-formyl receptor to actin nucleation in permeabilized human neutrophils. J Cell Biol 150:785-796

Goldschmidt-Clermont PJ, Machesky LM, Baldassare JJ, Pollard TD (1990) The actin-binding protein profilin binds to PIP2 and inhibits its hydrolysis by phospholipase C. Science 247:1575-1578

Goldschmidt-Clermont PJ, Kim JW, Machesky LM, Rhee SG, Pollard TD (1991) Regulation of phospholipase C-gamma 1 by profilin and tyrosine phosphorylation. Science 251:1231-1233

Goode BL, Drubin DG, Lappalainen P (1998) Regulation of the cortical actin cytoskeleton in budding yeast by twinfilin, a ubiquitous actin monomer-sequestering protein. J Cell Biol 142:723-733

Gray A, Van Der Kaay J, Downes CP (1999) The pleckstrin homology domains of protein kinase B and GRP1 (general receptor for phosphoinositides-1) are sensitive and selective probes for the cellular detection of phosphatidylinositol 3,4-bisphosphate and/or phosphatidylinositol 3,4,5-trisphosphate in vivo. Biochem J 344:929-936

Greenwood JA, Theibert AB, Prestwich GD, Murphy-Ullrich JE (2000) Restructuring of focal adhesion plaques by PI 3-kinase. Regulation by PtdIns (3,4,5)-p(3) binding to alpha-actinin. J Cell Biol 150:627-642

Gungabissoon RA, Jiang C-J, Drøbak BK, Maciver SK, Hussey P (1998) Interaction of maize-actin-depolymerising factor with actin and phosphoinositides and its inhibition of plant phospholipase C. Plant J 16:689-69

Hall A (1998) Rho GTPases and the actin cytoskeleton. Science 279:509-514

Hamada K, Shimizu T, Matsui T, Tsukita S, Hakoshima T (2000) Structural basis of the membrane-targeting and unmasking mechanisms of the radixin FERM domain. EMBO J 19:4449-4462

Han J, Luby-Phelps K, Das B, Shu X, Xia Y, Mosteller RD, Krishna UM, Falck JR, White MA, Broek D (1998) Role of substrates and products of PI 3-kinase in regulating activation of Rac-related guanosine triphosphatases by Vav. Science 279:558-560

Han X, Li G, Lin K (1997) Interactions between smooth muscle alpha-actinin and lipid bilayers. Biochem 36:10364-10371

Hart MC, Cooper JA (1999) Vertebrate isoforms of actin capping protein beta have distinct functions In vivo. J Cell Biol 147:1287-1298

Hartwig JH, Bokoch GM, Carpenter CL, Janmey PA, Taylor LA, Toker A, Stossel TP (1995) Thrombin receptor ligation and activated Rac uncap actin filament barbed ends through phosphoinositide synthesis in permeabilized human platelets. Cell 82:643-653

Haslam RJ, Koide HB, Hemmings BA (1993) Pleckstrin domain homology. Nature 363:309-310

Haugh JM, Codazzi F, Teruel M, Meyer T (2000) Spatial sensing in fibroblasts mediated by 3' phosphoinositides. J Cell Biol 151:1269-1280

Hawkins PT, Eguinoa A, Qiu RG, Stokoe D, Cooke FT, Walters R, Wennström S, Claesson-Welsh L, Evans T, Symons M (1995) PDGF stimulates an increase in GTP-Rac via activation of phosphoinositide 3-kinase. Curr Biol 5:393-403

He H, Watanabe T, Zhan X, Huang C, Schuuring E, Fukami K, Takenawa T, Kumar CC, Simpson RJ, Maruta H (1998) Role of phosphatidylinositol 4,5-bisphosphate in Ras/Rac-induced disruption of the cortactin-actomyosin II complex and malignant transformation. Mol Cell Biol 18:3829-3837

Higgs HN, Pollard TD (2000) Activation by Cdc42 and PIP(2) of Wiskott-Aldrich syndrome protein (WASp) stimulates actin nucleation by Arp2/3 complex. J Cell Biol 150:1311-1320

Hilpelä P, Oberbanscheidt P, Hahne P, Hund M, Kalhammer G, Small JV, Bähler M (2003) SWAP-70 identifies a transitional subset of actin filaments in motile cells. Mol Biol Cell (in press)

Hirao M, Sato N, Kondo T, Yonemura S, Monden M, Sasaki T, Takai Y, Tsukita S (1996) Regulation mechanism of ERM (ezrin/radixin/moesin) protein/plasma membrane association: possible involvement of phosphatidylinositol turnover and Rho-dependent signaling pathway. J Cell Biol 135:37-51

Hirose M, Ishizaki T, Watanabe N, Uehata M, Kranenburg O, Moolenaar WH, Matsumura F, Maekawa M, Bito H, Narumiya S (1998) Molecular dissection of the Rho-associated protein kinase (p160ROCK)-regulated neurite remodeling in neuroblastoma N1E-115 cells. J Cell Biol 141:1625-1636

Hirsch E, Katanaev VL, Garlanda C, Azzolino O, Pirola L, Silengo L, Sozzani S, Mantovani A, Altruda F, Wymann MP (2000) Central role for G protein-coupled phosphoinositide 3-kinase gamma in inflammation. Science 287:1049-1053

Homma K, Saito J, Ikebe R, Ikebe M (2001) Motor function and regulation of myosin X. J Biol Chem 276:34348-34354

Honda A, Nogami M, Yokozeki T, Yamazaki M, Nakamura H, Watanabe H, Kawamoto K, Nakayama K, Morris AJ, Frohman MA, Kanaho Y (1999) Phosphatidylinositol 4-phosphate 5-kinase alpha is a downstream effector of the small G protein ARF6 in membrane ruffle formation. Cell 99:521-532

Hopmann R, Cooper JA, Miller KG (1996) Actin organization, bristle morphology, and viability are affected by actin capping protein mutations in *Drosophila*. J Cell Biol 133:1293-1305

Insall RH, Weiner OD (2001) PIP3, PIP2, and cell movement—similar messages, different meanings? Dev Cell 1:743-747

Isakoff SJ, Cardozo T, Andreev J, Li Z, Ferguson KM, Abagyan R, Lemmon MA, Aronheim A, Skolnik EY (1998) Identification and analysis of PH domain-containing targets of phosphatidylinositol 3-kinase using a novel in vivo assay in yeast. EMBO J 17:5374-5387

Ishihara H, Shibasaki Y, Kizuki N, Katagiri H, Yazaki Y, Asano T, Oka Y (1996) Cloning of cDNAs encoding two isoforms of 68-kDa type I phosphatidylinositol-4-phosphate 5-kinase. J Biol Chem 271:23611-23614

Ishihara H, Shibasaki Y, Kizuki N, Wada T, Yazaki Y, Asano T, Oka Y (1998) Type I phosphatidylinositol-4-phosphate 5-kinases. Cloning of the third isoform and deletion/substitution analysis of members of this novel lipid kinase family. J Biol Chem 273:8741-8748

Ishihara H, Sasaoka T, Hori H, Wada T, Hirai H, Haruta T, Langlois WJ, Kobayashi M (1999) Molecular cloning of rat SH2-containing inositol phosphatase 2 (SHIP2) and its role in the regulation of insulin signaling. Biochem Biophys Res Commun 260:265-272

Janmey PA, Stossel TP, Allen PG (1987) Modulation of gelsolin function by phosphatidylinositol 4,5-bisphosphate. Nature 325:362–364

Janmey PA, Lamb J, Allen PG, Matsudaira PT (1992) Phosphoinositide-binding peptides derived from the sequences of gelsolin and villin. J Biol Chem 267:11818–11823

Jones DH, Morris JB, Morgan CP, Kondo H, Irvine RF, Cockcroft S (2000) Type I phosphatidylinositol 4-phosphate 5-kinase directly interacts with ADP-ribosylation factor 1 and is responsible for phosphatidylinositol 4,5-bisphosphate synthesis in the golgi compartment. J Biol Chem 275:13962–13966

Kavran JM, Klein DE, Lee A, Falasca M, Isakoff SJ, Skolnik EY, Lemmon MA (1998) Specificity and promiscuity in phosphoinositide binding by pleckstrin homology domains. J Biol Chem 273:30497–30508

Keely PJ, Westwick JK, Whitehead IP, Der CJ, Parise LV (1997) Cdc42 and Rac1 induce integrin-mediated cell motility and invasiveness through PI(3)K. Nature 390:632–636

Kim AS, Kakalis LT, Abdul-Manan N, Liu GA, Rosen MK (2000) Autoinhibition and activation mechanisms of the Wiskott-Aldrich syndrome protein. Nature 404:151–158

Kim J, Shishido T, Jiang X, Aderem A, McLaughlin S (1994) Phosphorylation, high ionic strength, and calmodulin reverse the binding of MARCKS to phospholipid vesicles. J Biol Chem 269:28214–28219

Kraulis PJ (1991) MOLSCRIPT: A program to produce both detailed and schematic plots of protein structures. J Appl Crystallogr 24:946–950

Kwiatkowski DJ (1999) Functions of gelsolin: motility, signaling, apoptosis, cancer. Curr Opin Cell Biol 11:103–108

Lambrechts A, Verschelde JL, Jonckheere V, Goethals M, Vandekerckhove J, Ampe C (1997) The mammalian profilin isoforms display complementary affinities for PIP2 and proline-rich sequences. EMBO J 16:484–494

Lambrechts A, Braun A, Jonckheere V, Aszodi A, Lanier LM, Robbens J, Van C, I, Vandekerckhove J, Fassler R, Ampe C (2000) Profilin II is alternatively spliced, resulting in profilin isoforms that are differentially expressed and have distinct biochemical properties. Mol Cell Biol 20:8209–8219

Lambrechts A, Jonckheere V, Dewitte D, Vandekerckhove J, Ampe C (2002) Mutational analysis of human profilin I reveals a second PI(4,5)-P2 binding site neighbouring the poly(L-proline) binding site. BMC Biochem 3:12

Lappalainen P, Fedorov EV, Fedorov AA, Almo SC, Drubin, DG (1997) Essential functions and actin-binding surfaces of yeast cofilin revealed by systematic mutagenesis. EMBO J 16:5520–5530

Lassing I, Lindberg U (1985) Specific interaction between phosphatidylinositol 4,5-bisphosphate and profilactin. Nature 314:472–474

Laux T, Fukami K, Thelen M, Golub T, Frey D, Caroni P (2000) GAP43, MARCKS, and CAP23 modulate PI(4,5)P(2) at plasmalemmal rafts, and regulate cell cortex actin dynamics through a common mechanism. J Cell Biol 149:1455–1472

Lemmon MA, Ferguson KM (2000) Signal-dependent membrane targeting by pleckstrin homology (PH) domains. Biochem J 350:1–18

Leung T, Chen XQ, Tan I, Manser E, Lim L (1998) Myotonic dystrophy kinase-related Cdc42-binding kinase acts as a Cdc42 effector in promoting cytoskeletal reorganization. Mol Cell Biol 18:130–140

Li Z, Jiang H, Xie W, Zhang Z, Smrcka AV, Wu D (2000) Roles of PLC-beta2 and -beta3 and PI3Kgamma in chemoattractant-mediated signal transduction. Science 287:1046–1049

Lin KM, Wenegieme E, Lu PJ, Chen CS, Yin HL (1997) Gelsolin binding to phosphatidylinositol 4,5-bisphosphate is modulated by calcium and pH. J Biol Chem 272:20443–20450

Ling K, Doughman RL, Firestone AJ, Bunce MW, Anderson RA (2002) Type Igamma phosphatidylinositol phosphate kinase targets and regulates focal adhesions. Nature 420:89–93

Lioubin MN, Algate PA, Tsai S, Carlberg K, Aebersold A, Rohrschneider LR (1996) p150Ship, a signal transduction molecule with inositol polyphosphate-5-phosphatase activity. Genes Dev 10:1084–1095

Liu BP, Burridge K (2000) Vav2 activates Rac1, Cdc42, and RhoA downstream from growth factor receptors but not beta1 integrins. Mol Cell Biol 20:7160–7169

Loijens JC, Anderson RA (1996) Type I phosphatidylinositol-4-phosphate 5-kinases are distinct members of this novel lipid kinase family. J Biol Chem 271:32937–32943

Lu PJ, Shieh WR, Rhee SG, Yin HL, Chen CS (1996) Lipid products of phosphoinositide 3-kinase bind human profilin with high affinity. Biochem 35:14027–14034

Ma L, Cantley LC, Janmey PA, Kirschner MW (1998) Corequirement of specific phosphoinositides and small GTP-binding protein Cdc42 in inducing actin assembly in *Xenopus* egg extracts. J Cell Biol 140:1125–1136

Machesky LM, Insall RH (1998) Scar1 and the related Wiskott-Aldrich syndrome protein, WASP, regulate the actin cytoskeleton through the Arp2/3 complex. Curr Biol 8:1347–1356

Machesky LM, Goldschmidt-Clermont PJ, Pollard TD (1990) The affinities of human platelet and *Acanthamoeba* profilin isoforms for polyphosphoinositides account for their relative abilities to inhibit phospholipase C. Cell Regulation 1:937–950

Maciver SK, Weeds AG (1994) Actophorin preferentially binds monomeric ADP-actin over ATP-bound actin: consequences for cell locomotion. FEBS Lett 347:251–256

Maehama T, Taylor GS, Dixon JE (2001) PTEN and myotubularin: novel phosphoinositide phosphatases. Annu Rev Biochem 70:247–279

Maekawa M, Ishizaki T, Boku S, Watanabe N, Fujita A, Iwamatsu A, Obinata T, Ohashi K, Mizuno K, Narumiya S (1999) Signaling from Rho to the actin cytoskeleton through protein kinases ROCK and LIM-kinase. Science 285:895–898

Marshall AJ, Krahn AK, Ma K, Duronio V, Hou S (2002) TAPP1 and TAPP2 are targets of phosphatidylinositol 3-kinase signaling in B cells: sustained plasma membrane recruitment triggered by the B-cell antigen receptor. Mol Cell Biol 22:5479–5491

Martel V, Racaud-Sultan C, Dupe S, Marie C, Paulhe F, Galmiche A, Block MR, Albiges-Rizo C (2001) Conformation, localization, and integrin binding of talin depend on its interaction with phosphoinositides. J Biol Chem 276:21217–21227

Martin TF (1998) Phosphoinositide lipids as signaling molecules: common themes for signal transduction, cytoskeletal regulation, and membrane trafficking. Annu Rev Cell Dev Biol 14:231–264

Matsui T, Yonemura S, Tsukita S (1999) Activation of ERM proteins in vivo by Rho involves phosphatidyl-inositol 4-phosphate 5-kinase and not ROCK kinases. Curr Biol 9:1259–1262

Mayer BJ, Ren R, Clark KL, Baltimore D (1993) A putative modular domain present in diverse signaling proteins. Cell 73:629–630

McLaughlin PJ, Gooch JT, Mannherz HG, Weeds AG (1993) Structure of gelsolin segment 1-actin complex and the mechanism of filament severing. Nature 364:685–692

McLaughlin S, Wang J, Gambhir A, Murray D (2002) PIP_2 and proteins: interactions, organization, and information flow. Annu Rev Biophys Biomol Struct 31:151–175

Meerschaert K, De C, V, De Ville Y, Vandekerckhove J, Gettemans J (1998) Gelsolin and functionally similar actin-binding proteins are regulated by lysophosphatidic acid. EMBO J 17:5923–5932

Meili R, Ellsworth C, Lee S, Reddy TB, Ma H, Firtel RA (1999) Chemoattractant-mediated transient activation and membrane localization of Akt/PKB is required for efficient chemotaxis to cAMP in *Dictyostelium*. EMBO J 18:2092–2105

Merrit EA, Murphy, Michael EP (1994) Raster3D Version 2.0: A program for photorealistic molecular graphics. Acta Crystallogr Sect D Biol Crystallogr D50:873

Michiels F, Stam JC, Hordijk PL, van der Kammen RA, Ruuls-Van SL, Feltkamp CA, Collard JG (1997) Regulated membrane localization of Tiam1, mediated by the NH_2-terminal pleckstrin homology domain, is required for Rac-dependent membrane ruffling and C-Jun NH_2-terminal kinase activation. J Cell Biol 137:387–398

Miki H, Takenawa T (1998) Direct binding of the verprolin-homology domain in N-WASP to actin is essential for cytoskeletal reorganization. Biochem Biophys Res Commun 243:73–78

Moriyama K, Yonezawa N, Sakai H, Yahara I, Nishida E (1992) Mutational analysis of an actin-binding site of cofilin and characterization of chimeric proteins between cofilin and destrin. J Biol Chem 267:7240–7244

Movilla N, Bustelo XR (1999) Biological and regulatory properties of Vav-3, a new member of the Vav family of oncoproteins. Mol Cell Biol 19:7870–7885

Mullins RD, Heuser JA, Pollard TD (1998) The interaction of Arp2/3 complex with actin: nucleation, high affinity pointed end capping, and formation of branching networks of filaments. Proc Natl Acad Sci USA 95:6181–6186

Murthy A, Gonzalez-Agosti C, Cordero E, Pinney D, Candia C, Solomon F, Gusella J, Ramesh V (1998) NHE-RF, a regulatory cofactor for Na^+-H^+ exchange, is a common interactor for merlin and ERM (MERM) proteins. J Biol Chem 273:1273–1276

Musacchio A, Gibson T, Rice P, Thompson J, Saraste M (1993) The PH domain: a common piece in the structural patchwork of signalling proteins. Trends Biochem Sci 18:343–348

Nakamura F, Huang L, Pestonjamasp K, Luna EJ, Furthmayr H (1999) Regulation of F-actin binding to platelet moesin in vitro by both phosphorylation of threonine 558 and polyphosphatidylinositides. Mol Biol Cell 10:2669–2685

Niggli V (2000) A membrane-permeant ester of phosphatidylinositol 3,4, 5-trisphosphate (PIP$_3$) is an activator of human neutrophil migration. FEBS Lett 473:217–221

Niggli V (2001) Structural properties of lipid-binding sites in cytoskeletal proteins. Trends Biochem Sci 26:604–611

Nobes CD, Hall A (1995) Rho, rac, and cdc42 GTPases regulate the assembly of multimolecular focal complexes associated with actin stress fibers, lamellipodia, and filopodia. Cell 81:53–62

Ohmori S, Sakai N, Shirai Y, Yamamoto H, Miyamoto E, Shimizu N, Saito N (2000) Importance of protein kinase C targeting for the phosphorylation of its substrate, myristoylated alanine-rich C-kinase substrate. J Biol Chem 275:26449–26457

Ojala PJ, Paavilainen V, Lappalainen P (2001) Identification of yeast cofilin residues specific for actin monomer and PIP2 binding. Biochemistry 40:15562–15569

Olave IA, Reck-Peterson SL, Crabtree GR (2002) Nuclear actin and actin-related proteins in chromatin remodeling. Annu Rev Biochem 71:755–781

Palmgren S, Ojala PJ, Wear MA, Cooper JA, Lappalainen P (2001) Interactions with PIP2, ADP-actin monomers, and capping protein regulate the activity and localization of yeast twinfilin. J Cell Biol 155:251–260

Palmgren S, Vartiainen M, Lappalainen P (2002) Twinfilin, a molecular mailman for actin monomers. J Cell Sci 115:881–886

Panebra A, Ma SX, Zhai LW, Wang XT, Rhee SG, Khurana S (2001) Regulation of phospholipase C-γ_1 by the actin-regulatory protein villin. Am J Physiol Cell Physiol 281:C1046–C1058

Pantaloni D, Carlier MF (1993) How profilin promotes actin filament assembly in the presence of thymosin beta 4. Cell 75:1007–1014

Pearson MA, Reczek D, Bretscher A, Karplus PA (2000) Structure of the ERM protein moesin reveals the FERM domain fold masked by an extended actin binding tail domain. Cell 101:259–270

Pollard TD, Blanchoin L, Mullins RD (2000) Molecular mechanisms controlling actin filament dynamics in nonmuscle cells. Annu Rev Biophys Biomol Struct 29:545–576

Prehoda KE, Scott JA, Mullins RD, Lim WA (2000) Integration of multiple signals through cooperative regulation of the N-WASP-Arp2/3 complex. Science 290:801–806

Pruyne D, Evangelista M, Yang C, Bi E, Zigmond S, Bretscher A, Boone C (2002) Role of formins in actin assembly: nucleation and barbed-end association. Science 297:612–615

Raghunathan V, Mowery P, Rozycki M, Lindberg U, Schutt (1992) Structural changes in profilin accompany its binding to phosphatidylinositol, 4,5-bisphosphate. FEBS Lett 297:46–50

Rameh LE, Cantley LC (1999) The role of phosphoinositide 3-kinase lipid products in cell function. J Biol Chem 274:8347–8350

Rameh LE, Arvidsson A, Carraway KL, Couvillon AD, Rathbun G, Crompton A, Van-Renterghem B, Czech MP, Ravichandran KS, Burakoff SJ, Wang DS, Chen CS, Cantley LC (1997) A comparative analysis of the phosphoinositide binding specificity of pleckstrin homology domains. J Biol Chem 272:22059–22066

Raucher D, Stauffer T, Chen W, Shen K, Guo S, York JD, Sheetz MP, Meyer T (2000) Phosphatidylinositol 4,5-bisphosphate functions as a second messenger that regulates cytoskeleton-plasma membrane adhesion. Cell 100:221–228

Reczek D, Berryman M, Bretscher A (1997) Identification of EBP50: A PDZ-containing phosphoprotein that associates with members of the ezrin-radixin-moesin family. J Cell Biol 139:169–179

Rees DJ, Ades SE, Singer SJ, Hynes RO (1990) Sequence and domain structure of talin. Nature 347:685–689

Ren XD, Schwartz MA (1998) Regulation of inositol lipid kinases by Rho and Rac. Curr Opin Genet Dev 8:63–67

Ren XD, Bokoch GM, Traynor-Kaplan A, Jenkins GH, Anderson RA, Schwartz MA (1996) Physical association of the small GTPase Rho with a 68-kDa phosphatidylinositol 4-phosphate 5-kinase in Swiss 3T3 cells. Mol Biol Cell 7:435–442

Rickert P, Weiner OD, Wang F, Bourne HR, Servant G (2000) Leukocytes navigate by compass: roles of PI3Kgamma and its lipid products. Trends Cell Biol 10:466–473

Ridley AJ, Hall A (1992) The small GTP-binding protein rho regulates the assembly of focal adhesions and actin stress fibers in response to growth factors. Cell 70:389–399

Ridley AJ, Paterson HF, Johnston CL, Diekmann D, Hall A (1992) The small GTP-binding protein rac regulates growth factor-induced membrane ruffling. Cell 70:401–410

Rodal AA, Tetreault JW, Lappalainen P, Drubin DG, Amberg DC (1999) Aip1p interacts with cofilin to disassemble actin filaments. J Cell Biol 145:1251–1264

Rohatgi R, Ma L, Miki H, Lopez M, Kirchhausen T, Takenawa T, Kirschner MW (1999) The interaction between N-WASP and the Arp2/3 complex links Cdc42-dependent signals to actin assembly. Cell 97:221–231

Rohatgi R, Ho HY, Kirschner MW (2000) Mechanism of N-WASP activation by CDC42 and phosphatidylinositol 4,5-bisphosphate. J Cell Biol 150:1299–1310

Rohatgi R, Nollau P, Ho HY, Kirschner MW, Mayer BJ (2001) Nck and phosphatidylinositol 4,5-bisphosphate synergistically activate actin polymerization through the N-WASP-Arp2/3 pathway. J Biol Chem 276:26448–26452

Rozelle AL, Machesky LM, Yamamoto M, Driessens MH, Insall RH, Roth MG, Luby-Phelps K, Marriott G, Hall A, Yin HL (2000) Phosphatidylinositol 4,5-bisphosphate induces actin-based movement of raft-enriched vesicles through WASP-Arp2/3. Curr Biol 10:311–320

Russo C, Gao Y, Mancini P, Vanni C, Porotto M, Falasca M, Torrisi MR, Zheng Y, Eva A (2001) Modulation of oncogenic DBL activity by phosphoinositol phosphate binding to pleckstrin homology domain. J Biol Chem 276:19524–19531

Sagot I, Klee SK, Pellman D (2002a) Yeast formins regulate cell polarity by controlling the assembly of actin cables. Nat Cell Biol 4:42–50

Sagot I, Rodal AA, Moseley J, Goode BL, Pellman D (2002b) An actin nucleation mechanism mediated by Bni1 and profilin. Nat Cell Biol 4:626–631

Sakisaka T, Itoh T, Miura K, Takenawa T (1997) Phosphatidylinositol 4,5-bisphosphate phosphatase regulates the rearrangement of actin filaments. Mol Cell Biol 17:3841–3849

Sander EE, van Delft S, ten Klooster JP, Reid T, van der Kammen RA, Michiels F, Collard JG (1998) Matrix-dependent Tiam1/Rac signaling in epithelial cells promotes

either cell-cell adhesion or cell migration and is regulated by phosphatidylinositol 3-kinase. J Cell Biol 143:1385-1398
Sanders LC, Matsumura F, Bokoch GM, de Lanerolle P (1999) Inhibition of myosin light chain kinase by p21-activated kinase. Science 283:2083-2085
Sasaki T, Irie-Sasaki J, Jones RG, Oliveira dSA, Stanford WL, Bolon B, Wakeham A, Itie A, Bouchard D, Kozieradzki I, Joza N, Mak TW, Ohashi PS, Suzuki A, Penninger JM (2000) Function of PI3Kgamma in thymocyte development, T cell activation, and neutrophil migration. Science 287:1040-1046
Schafer DA, Korshunova YO, Schroer TA, Cooper JA (1994) Differential localization and sequence analysis of capping protein beta-subunit isoforms of vertebrates. J Cell Biol 127:453-465
Schafer DA, Jennings PB, Cooper JA (1996) Dynamics of capping protein and actin assembly in vitro: uncapping barbed ends by polyphosphoinositides. J Cell Biol 135:169-179
Schluter K, Jockusch BM, Rothkegel M (1997) Profilins as regulators of actin dynamics. Biochim Biophys Acta 1359:97-109
Schuebel KE, Movilla N, Rosa JL, Bustelo XR (1998) Phosphorylation-dependent and constitutive activation of Rho proteins by wild-type and oncogenic Vav-2. EMBO J 17:6608-6621
Sellers JR (2000) Myosins: a diverse superfamily. Biochim Biophys Acta 1496:3-22
Serunian LA, Auger KR, Roberts TM, Cantley LC (1990) Production of novel polyphosphoinositides in vivo is linked to cell transformation by polyomavirus middle T antigen. J Virol 64:4718-4725
Servant G, Weiner OD, Herzmark P, Balla T, Sedat JW, Bourne HR (2000) Polarization of chemoattractant receptor signaling during neutrophil chemotaxis. Science 287:1037-1040
Shaw LM, Rabinovitz I, Wang HH, Toker A, Mercurio AM (1997) Activation of phosphoinositide 3-OH kinase by the alpha6beta4 integrin promotes carcinoma invasion. Cell 91:949-960
Sheterline P, Clayton J, Sparrow J (1998) Actin. Protein Profile 4:1-119
Shibasaki F, Fukami K, Fukui Y, Takenawa T (1994) Phosphatidylinositol 3-kinase binds to alpha-actinin through the p85 subunit. Biochem J 302:551-557
Shibasaki Y, Ishihara H, Kizuki N, Asano T, Oka Y, Yazaki Y (1997) Massive actin polymerization induced by phosphatidylinositol-4-phosphate 5-kinase in vivo. J Biol Chem 272:7578-7581
Short DB, Trotter KW, Reczek D, Kreda SM, Bretscher A, Boucher RC, Stutts MJ, Milgram SL (1998) An apical PDZ protein anchors the cystic fibrosis transmembrane conductance regulator to the cytoskeleton. J Biol Chem 273:19797-19801
Simson R, Wallraff E, Faix J, Niewohner J, Gerisch G, Sackmann E (1998) Membrane bending modulus and adhesion energy of wild-type and mutant cells of Dictyostelium lacking talin or cortexillins. Biophys J 74:514-522
Singh SS, Chauhan A, Murakami N, Chauhan VP (1996) Profilin and gelsolin stimulate phosphatidylinositol 3-kinase activity. Biochem 35:16544-16549
Skare P, Karlsson R (2002) Evidence for two interaction regions for phosphatidylinositol(4,5)-bisphosphate on mammalian profilin I. FEBS Lett 522:119-124

Snyder JT, Rossman KL, Baumeister MA, Pruitt WM, Siderowski DP, Der CJ, Lemmon MA, Sondek J (2001) Quantitative analysis of the effect of phosphoinositide interactions on the function of Dbl family proteins. J Biol Chem 276:45868–45875

Southwick FS, DiNubile MJ (1986) Rabbit alveolar macrophages contain a Ca^{2+}-sensitive, 41,000-dalton protein which reversibly blocks the "barbed" ends of actin filaments but does not sever them. J Biol Chem 261:14191–14195

Stauffer TP, Ahn S, Meyer T (1998) Receptor-induced transient reduction in plasma membrane PtdIns(4,5)P2 concentration monitored in living cells. Curr Biol 8:343–346

Steimle PA, Hoffert JD, Adey NB, Craig SW (1999) Polyphosphoinositides inhibit the interaction of vinculin with actin filaments. J Biol Chem 274:18414–18420

Stephens LR, Hughes KT, Irvine RF (1991) Pathway of phosphatidylinositol(3,4,5)-trisphosphate synthesis in activated neutrophils. Nature 351:33–39

Stephens LR, Jackson TR, Hawkins PT (1993) Agonist-stimulated synthesis of phosphatidylinositol(3,4,5)-trisphosphate: a new intracellular signalling system? Biochim Biophys Acta 1179:27–75

Stock A, Steinmetz MO, Janmey PA, Aebi U, Gerisch G, Kammerer RA, Weber I, Faix J (1999) Domain analysis of cortexillin I: actin-bundling, PIP_2-binding and the rescue of cytokinesis. EMBO J 18:5274–5284

Stossel TP, Condeelis J, Cooley L, Hartwig JH, Noegel A, Schleicher M, Shapiro SS (2001) Filamins as integrators of cell mechanics and signalling. Nat Rev Mol Cell Biol 2:138–145

Sumi T, Matsumoto K, Shibuya A, Nakamura T (2001) Activation of LIM kinases by myotonic dystrophy kinase-related Cdc42-binding kinase alpha. J Biol Chem 276:23092–23096

Sun H, Lin K, Yin HL (1997) Gelsolin modulates phospholipase C activity in vivo through phospholipid binding. J Cell Biol 138:811–820

Sun HQ, Kwiatkowska K, Wooten DC, Yin HL (1995) Effects of CapG overexpression on agonist-induced motility and second messenger generation. J Cell Biol 129:147–156

Sun HQ, Yamamoto M, Mejillano M, Yin HL (1999) Gelsolin, a multifunctional actin regulatory protein. J Biol Chem 274:33179–33182

Tall EG, Spector I, Pentyala SN, Bitter I, Rebecchi MJ (2000) Dynamics of phosphatidylinositol 4,5-bisphosphate in actin-rich structures. Curr Biol 10:743–746

Tempel M, Goldmann WH, Dietrich C, Niggli V, Weber T, Sackmann E, Isenberg G (1994) Insertion of filamin into lipid membranes examined by calorimetry, the film balance technique, and lipid photolabeling. Biochem 33:12565–12572

Thomas CC, Dowler S, Deak M, Alessi DR, van Aalten DM (2001) Crystal structure of the phosphatidylinositol 3,4-bisphosphate-binding pleckstrin homology (PH) domain of tandem PH-domain-containing protein 1 (TAPP1): molecular basis of lipid specificity. Biochem J 358:287–294

Thrasher AJ (2002) WASp in immune-system organization and function. Nat Rev Immunol 2:635–646

Tolias K, Carpenter CL (2000) In vitro interaction of phosphoinositide-4-phosphate 5-kinases with Rac. Methods Enzymol 325:190–200

Tolias KF, Hartwig JH, Ishihara H, Shibasaki Y, Cantley LC, Carpenter CL (2000) Type Ialpha phosphatidylinositol-4-phosphate 5-kinase mediates Rac-dependent actin assembly. Curr Biol 10:153-156

Tuominen EK, Holopainen JM, Chen J, Prestwich GD, Bachiller PR, Kinnunen, PK, Janmey PA (1999) Fluorescent phosphoinositide derivatives reveal specific binding of gelsolin and other actin regulatory proteins to mixed lipid bilayers. Eur J Biochem 263:85-92

Van Aelst L, D'Souza-Schorey C (1997) Rho GTPases and signaling networks. Genes Dev 11:2295-2322

Van Aelst L, Joneson T, Bar-Sagi D (1996) Identification of a novel Rac1-interacting protein involved in membrane ruffling. EMBO J 15:3778-3786

van der Flier A, Sonnenberg A (2001) Structural and functional aspects of filamins. Biochim Biophys Acta 1538:99-117

Van Rheenen J, Jalink K (2002) Agonist-induced PIP_2 hydrolysis inhibits cortical actin dynamics: Regulation at a global but not at a micrometer scale. Mol Biol Cell 13:3257-3267

Van Troys M, Dewitte D, Verschelde JL, Goethals M, Vandekerckhove J, Ampe C (2000) The competitive interaction of actin and PIP2 with actophorin is based on overlapping target sites: design of a gain-of-function mutant. Biochem 39:12181-12189

Vartiainen M, Ojala PJ, Auvinen P, Peränen J, Lappalainen P (2000) Mouse A6/twinfilin is an actin monomer-binding protein that localizes to the regions of rapid actin dynamics. Mol Cell Biol 20:1772-1783

Varticovski L, Druker B, Morrison D, Cantley L, Roberts T (1989) The colony stimulating factor-1 receptor associates with and activates phosphatidylinositol-3 kinase. Nature 342:699-702

Varticovski L, Daley GQ, Jackson P, Baltimore D, Cantley LC (1991) Activation of phosphatidylinositol 3-kinase in cells expressing abl oncogene variants. Mol Cell Biol 11:1107-1113

Volberg T, Geiger B, Kam Z, Pankov R, Simcha I, Sabanay H, Coll JL, Adamson E, Ben-Ze'ev A (1995) Focal adhesion formation by F9 embryonal carcinoma cells after vinculin gene disruption. J Cell Sci 108:2253-2260

Wahlström G, Vartiainen M, Yamamoto L, Mattila PK, Lappalainen P, Heino TI (2001) Twinfilin is required for actin-dependent developmental processes in Drosophila. J Cell Biol 155:787-796

Wang J, Arbuzova A, Hangyas-Mihalyne G, McLaughlin S (2001) The effector domain of myristoylated alanine-rich C kinase substrate binds strongly to phosphatidylinositol 4,5-bisphosphate. J Biol Chem 276:5012-5019

Watanabe N, Madaule P, Reid T, Ishizaki T, Watanabe G, Kakizuka A, Saito Y, Nakao K, Jockusch BM, Narumiya S (1997) p140mDia, a mammalian homolog of Drosophila diaphanous, is a target protein for Rho small GTPase and is a ligand for profilin. EMBO J 16:3044-3056

Watt SA, Kular G, Fleming IN, Downes CP, Lucocq JM (2002) Subcellular localization of phosphatidylinositol 4,5-bisphosphate using the pleckstrin homology domain of phospholipase C delta1. Biochem J 363:657-666

Weber I, Gerisch G, Heizer C, Murphy J, Badelt K, Stock A, Schwartz JM, Faix J (1999) Cytokinesis mediated through the recruitment of cortexillins into the cleavage furrow. EMBO J 18:586–594

Weed SA, Parsons JT (2001) Cortactin: coupling membrane dynamics to cortical actin assembly. Oncogene 20:6418–6434

Weeds A, Maciver S (1993) F-actin capping proteins. Curr Opin Cell Biol 5:63–69

Welch MD, Mullins RD (2002) Cellular control of actin nucleation. Annu Rev Cell Dev Biol 18:247–288

Wennström S, Hawkins P, Cooke F, Hara K, Yonezawa K, Kasuga M, Jackson T, Claesson-Welsh L, Stephens L (1994a) Activation of phosphoinositide 3-kinase is required for PDGF-stimulated membrane ruffling. Curr Biol 4:385–393

Wennström S, Siegbahn A, Yokote K, Arvidsson AK, Heldin CH, Mori S, Claesson-Welsh L (1994b) Membrane ruffling and chemotaxis transduced by the PDGF beta-receptor require the binding site for phosphatidylinositol 3′ kinase. Oncogene 9:651–660

Wymann M, Arcaro A (1994) Platelet-derived growth factor-induced phosphatidylinositol 3-kinase activation mediates actin rearrangements in fibroblasts. Biochem J 298:517–520

Xian W, Janmey PA (2002) Dissecting the gelsolin-polyphosphoinositide interaction and engineering of a polyphosphoinositide-sensitive gelsolin C-terminal half protein. J Mol Biol 322:755

Yamamoto M, Hilgemann DH, Feng S, Bito H, Ishihara H, Shibasaki Y, Yin HL (2001) Phosphatidylinositol 4,5-bisphosphate induces actin stress-fiber formation and inhibits membrane ruffling in CV1 cells. J Cell Biol 152:867–876

Yonemura S, Matsui T, Tsukita S (2002) Rho-dependent and -independent activation mechanisms of ezrin/radixin/moesin proteins: an essential role for polyphosphoinositides in vivo. J Cell Sci 115:2569–2580

Yonezawa N, Nishida E, Iida K, Yahara I, Sakai H (1990) Inhibition of the interactions of cofilin, destrin, and deoxyribonuclease I with actin by phosphoinositides. J Biol Chem 265:8382–8386

Young P, Gautel M (2000) The interaction of titin and alpha-actinin is controlled by a phospholipid-regulated intramolecular pseudoligand mechanism. EMBO J 19:6331–6340

Yu FX, Sun HQ, Janmey PA, Yin HL (1992) Identification of a polyphosphoinositide-binding sequence in an actin monomer-binding domain of gelsolin. J Biol Chem 267:14616–14621

Yun CH, Lamprecht G, Forster DV, Sidor A (1998) NHE3 kinase A regulatory protein E3KARP binds the epithelial brush border Na^+/H^+ exchanger NHE3 and the cytoskeletal protein ezrin. J Biol Chem 273:25856–25863

Zeng L, Sachdev P, Yan L, Chan JL, Trenkle T, McClelland M, Welsh J, Wang LH (2000) Vav3 mediates receptor protein tyrosine kinase signaling, regulates GTPase activity, modulates cell morphology, and induces cell transformation. Mol Cell Biol 20:9212–9224

Roles of PI3K in Neutrophil Function

M. O. Hannigan[1] · C. K. Huang[2] · D. Q. Wu[1]

[1] Department of Genetics and Developmental Biology,
University of Connecticut Health Center, 263 Farmington Ave,
Farmington, CT 06032, USA
E-mail: dwu@neuron.uchc.edu

[2] Department of Pathology, University of Connecticut Health Center,
263 Farmington Ave, Farmington, CT 06032, USA

1	Introduction	166
2	Phosphoinositide 3-Kinases	166
3	PI3K in Chemotaxis	167
4	PI3K in Phagocytosis	169
4.1	Engulfment	169
4.2	Internalization	170
5	PI3K in Superoxide Production	170
5.1	Neutrophil Superoxide Production	170
5.2	Neutrophil Priming	171
6	Conclusion	171
References		172

Abstract Neutrophils are terminally differentiated cells that play a vital role in host defense. It has recently become evident that phospholipid regulation plays an import role in many neutrophil functions. We review the regulation of neutrophil functions such as chemotaxis, superoxide production, and phagocytosis by phosphatidylinositol-3,4,5-trisphosphate (PIP_3), which is generated in neutrophils by PI3Kγ. Several lines of evidence are presented demonstrating the importance of this kinase in regulating chemotaxis, in particular the directionality of chemotactic migration. Evidence suggesting that this kinase is important for phagocytosis, especially during engulfment and the internalization of large particles, is also reviewed. Finally, it is suggested that PI3K is important for superoxide production and neutrophil priming. The common link between these seemingly diverse functions is that PI3Kγ, via its phospholipid products, appears to be providing spatial-temporal cues for the binding of actin-organizing proteins.

1
Introduction

Neutrophils comprise an essential part of the innate immune response. They circulate in large numbers and migrate into tissues in response to inflammatory cues. This exodus from the vasculature to sites of inflammation requires the coordination of numerous cellular systems. The coordination is required to balance the need to combat invading organisms yet minimize the potential damage to healthy tissues.

The importance of neutrophils in the defense against bacterial infection is demonstrated by the severity of diseases that compromise neutrophil function (Roos and Law 2001). It is increasingly evident that phospholipid regulation, by various lipid kinases, phosphatases, and lipases, plays an important role in the functioning of neutrophils. In particular, phosphatidylinositol-3,4,5-trisphosphate (PIP_3) has been demonstrated to be important in such neutrophil functions as chemotaxis, superoxide production, and phagocytosis.

2
Phosphoinositide 3-Kinases

The phoshophoinositide 3-kinases (PI3K) are evolutionarily conserved lipid kinases composed of a catalytic subunit and a regulatory subunit that adds phosphate to the D3 position of phosphatidylinositol. These kinases are involved in numerous cellular functions and responses (reviewed in Wymann and Pirola 1998 and Wu et al. 2000). The structure of the different PI3K classes has been extensively reviewed (Fruman et al. 1998).

Type I PI3K are understood in the most detail with regards to leukocyte signaling (Fruman and Cantley 2002). Type I PI3K phosphorylates phosphatidylinositol 4,5-bisphosphate. The resulting $PI3,4,5P_3$ serves as a lipid ligand of PH domain-containing proteins (reviewed in Corvera 2001). Type I PI3K are further divided into type Ia, which are activated by tyrosine kinases, and type Ib, which are activated by the $\beta\gamma$ complex of heterotrimeric G proteins (Dekker and Segal 2000).

Type II PI3K, although poorly understood, are associated with and regulated by clathrin (Stephens et al. 2002). The type III PI3K produce PI3P, which is the lipid ligand for FYVE and PX domain-containing proteins and functions in endosomal membrane targeting (reviewed in Corvera 2001).

As with many biological systems, balance is key to the proper regulation of phospholipids in cells. In this case, the negative regulators of the PI3K are the phosphatases PTEN (for "phosphatase and tensin homolog deleted on chromosome 10"), and *SH2*-containing inositol phosphatase (SHIP) (reviewed in Krystal 2000).

3
PI3K in Chemotaxis

The first step in reaching an inflammation site requires neutrophils to leave the circulating pool and migrate across the endothelium. It has been demonstrated that PI3K signaling enhances the affinity of β_3 integrins for their ligands (Bruyninckx et al. 2001). This in effect slows the neutrophil and allows it to better interact with signals on the endothelial matrix. This is similar to what has been observed in vitro with a monocyte/endothelial cell system (Gerszten et al. 2001). After crossing the endothelial barrier chemotactic signals direct the neutrophils to sites of inflammation.

Many chemotactic receptors belong to the G protein-coupled receptor family and thus activate the type 1b class of PI3K after receptor engagement. At least in murine neutrophils it has been shown that the majority of PIP3 generated in response to *N*-formyl-Met-Leu-Phe peptide (*f*MLP) stimulation is due to the p110γ isoform of PI3K (Li et al. 2002). Additionally, in *f*MLP-stimulated neutrophils, PIP_3 was localized at the very leading edge of migrating cells (Rickert et al. 2000). This has led to the hypothesis that in *f*MLP-stimulated neutrophils, the function of PI3Kγ is to produce PIP_3 at the leading edge. This localized increase in PIP_3 may act much like a compass to control the direction of cell migration (Bourne and Weiner 2002).

PI3Kγ-knockout mice have been generated, and neutrophils from these mice have been examined for various responses to several stimuli. In the Boyden chamber chemotaxis assay PI3K$\gamma^{-/-}$ neutrophils showed impaired migration in response to IL-8 (Hirsh et al. 2000), *f*MLP (Li et al. 2000; Hirsh et al. 2000; Sasaki et al. 2000), C5a (Hirsh et al. 2000;

Sasaki et al. 2000), and MIP-1α (Li et al. 2000). Furthermore, time-lapse video microscopy of PI3Kγ$^{-/-}$ neutrophils in Zigmond chambers containing fMLP gradients demonstrated that these neutrophils are defective in migration directionality with lower cell motility, poor translocation, and frequent changes in direction (Hannigan et al. 2002).

It is hypothesized that directional control may occur through the binding of signaling molecules containing pleckstrin homology (PH) domains (such as the protein kinase AKT) to PIP$_3$ (Rappel et al. 2002). In support of this, when expressed in the neutrophil-like HL60 cells, PHAKT-GFP (the PH domain of AKT kinase tagged with GFP) shows selective recruitment to the cell regions receiving the strongest fMLP stimulation. These regions also showed F-actin polymerization and pseudopod formation (Servant et al. 2000). Thus the primary role of PI3K during chemotaxis seems to be in providing the binding partner for PH domain-containing proteins. In support of this, the PI3Kγ$^{-/-}$ neutrophils were found to be defective in their ability to stabilize and consolidate a leading edge, which likely leads to the inability to maintain cellular translocation in a persistent direction (Hannigan et al. 2002). Additionally, *Dictyostelium* mutants lacking *pi3k1/2* show the requirement for PIP$_3$ in the spatiotemporal regulation of F-actin and the establishment of pseudopods toward chemotactic gradients in this system (Funamoto et al. 2001).

The production of PIP$_3$ by PI3K at the leading edge as a result of chemotactic stimulation would recruit PH domain-containing proteins, such as AKT, to this region of the cell membrane, which in turn would cause the redistribution of the cellular machineries that provide directionality for cell movement (i.e., F-actin). Recent reports have shown that local PIP$_3$ levels, regulated by PI3K and the PI3 phosphatase PTEN are important for maintaining persistent cell polarity and directed migration in *Dictyostelium* (Funamoto et al. 2002; Iijima et al. 2002). The role of PTEN in neutrophil chemotaxis has not been directly investigated.

Some intriguing questions remain regarding the role of PI3K in neutrophil chemotaxis. Although PIP$_3$ provides a compass to follow the correct direction during migration, it is unclear how the initial polarizing signals are determined. In other words, before the morphological polarization observed during chemotaxis, how is the internal biochemical polarity determined? Additionally, the roles of other cellular factors in influencing PI3K activity are poorly understood. It has been suggested that positive and negative regulatory links between PIP_3 and Rho family GTPases constitute a broadly conserved module for the establishment of cell polarity during eukaryotic chemotaxis (Weiner et al. 2002).

4
PI3K in Phagocytosis

A vital function of neutrophils is to phagocytose opsonized particles such as invading bacteria. The complete process of phagocytosis occurs in two distinct phases, 1) engulfment and 2) internalization. Disruption of PI3K in *Dictyostelium* mutants severely impairs macropinocytosis, a process similar to phagocytosis (Zhuo et al. 1998). With gene-targeted deletion in mice it has also been shown that the class III PI3K are vital for phagolysosomal formation, whereas class I PI3K are important for phagolysosomal maturation (Viera et al. 2001).

4.1
Engulfment

Immediately after engagement of the neutrophil Fcγ receptors, F-actin polymerizes on the cytoplasmic face of the neutrophil membrane facing the phagocytic target. This very early step is dependent on tyrosine phosphorylation but seems to be independent of PI3K activity. However, soon after engagement, pseudopodia begin to engulf the phagocytic target in a PI3K-dependent process (Ninomiya et al. 1994). This involves the rearrangement and extension of actin filaments to cause membrane to surround the phagocytic target. In this case, the type 1a PI3K are activated by tyrosine phosphorylated adapter proteins. The non-receptor tyrosine kinase Syk is especially important as the protein kinase, which activates many of the adapter molecules (Majeed et al. 2001). The engulfment process is complete when the pseudopodia completely surround the phagocytic target and the membrane zippers itself shut. Use of the PI3K inhibitor wortmannin suggests that PI3K plays a role in activating p21-activated kinase 1 (PAK1) in this engulfment process (Dharmawardhane et al. 1999).

4.2
Internalization

The ingestion of larger particles seems to be more dependent on PI3K activity compared to smaller particles. This suggests another role of PI3K during phagocytosis. PI3K may assist in the recruiting of new membrane to the site of extending pseudopodia. It has been shown that new membrane for phagocytosis is derived from the fusion of vesicles with in the neutrophil (Perskvist et al. 2002). It is known that neutrophils treated with wortmannin show an inhibition in vesicle fusion (Detmers et al. 1996). More recently, the delivery of membrane has been linked to PI3K regulation of ADP ribosylation factor (ARF) guanosine nucleotide exchange factor and ARF GTPase-activating proteins (Krugman et al. 2002). In the final stage of engulfment, when the membranes seal around the particle, the role of PI3K is not clear. However, in a *Dictyostelium* model with similarities to neutrophils closure of the membranes occurred only after diminishment of PIP_3 levels (Rupper et al. 2001).

5
PI3K in Superoxide Production

5.1
Neutrophil Superoxide Production

The PI3K inhibitors wortmaninin and LY294002 have both been shown to inhibit superoxide production, suggesting a role for PI3K in this process (Yasui and Komiyama 2001). Additionally, during phagocytosis, it has been observed that the levels of PIP_3 rise in phagosomal membranes (Vieiri et al. 2001; Ellson et al. 2001a). This suggests that PIP_3 may be involved in membrane targeting of the NADPH oxidase.

To be fully functional, the NADPH oxidase complex requires the translocation of proteins $p47^{phox}$ $p67^{phox}$, and $p40^{phox}$ from the cytosol to the cell membrane (Vignais 2002). The $p47^{phox}$ and $p67^{phox}$ are activating subunits of the oxidase complex, whereas $p40^{phox}$ serves as an adapter molecule (Sato et al. 2001). The amino acid sequences of both $p40^{phox}$ and $p47^{phox}$ contain Phox homology (PX) domains (Yaffe 2002). The PX domain has been demonstrated to be a lipid binding domain (Song 2001, Hiroaki et al. 2001). It is the interaction of the PX domains of p40^{phox} and p47^{phox} that spatially and temporally target these proteins to appropriate subcellular membranes by PI3K (Kanai et al. 2001). Although the p67^{phox} component of the oxidase complex is recruited to this site by its binding to p40^{phox}, it is apparent that for proper stability all three proteins must be present (Noack et al. 2001).

The X-ray structure of p40^{phox} reveals that the PX domain has a fold distinctly different from other PIP$_3$ binding domains, such as PH and FYVE; however, like these more traditional domains, PX utilizes three basic residues to bind lipid headgroups (Bravo et al. 2001). The X-ray structure of the p47^{phox} PX domain was also recently resolved and surprisingly demonstrated to have two unique lipid binding sites, suggesting that membrane phospholipids play an important role in the regulation of superoxide production (Karathanassis et al. 2002).

5.2
Neutrophil Priming

PI3K has also been demonstrated to play an important role in neutrophil priming (Condliffe et al. 1988). Priming is the process by which neutrophils that have been exposed to various inflammatory mediators produce a much larger respiratory burst compared with unprimed neutrophils. In mouse neutrophils stimulated with TNF-α or GMCSF, levels of PIP3 found at 10–60 s after fMLP-treatment were increased (Cadwallader et al. 2002). In addition, the priming effect was abrogated by LY294002. This suggests that the timing and magnitude of PtdIns(3,4,5)P3 accumulation in neutrophils is closely correlated with O_2^- generation. Other experiments suggest that PI3Kγ is responsible for the enhanced PIP$_3$ production seen in primed cells, and that factors other than activation of p21(ras) underlie this response (Cadwallader et al. 2002). Mice lacking the p110γ subunit of PI3K are defective in their ability to produce superoxide (Li et al. 2000); however, priming has not been examined in these neutrophils.

6
Conclusion

The PI3K pathway regulates numerous cellular processes including proliferation, growth, apoptosis, and cytoskeletal rearrangement. As neutrophils are terminally differentiated, PI3K appears to be most important for providing spatiotemporal cues to architects of the actin cytoskeleton by providing binding sites for lipid binding domain-containing proteins.

Directional sensing is a fundamental mechanism of cells ranging from neutrophils to *Dictyostelium*. The study of neutrophil PI3K provides a model system for learning how cells can determine the orientation of chemical gradients. This pathway appears to be vital for neutrophil chemotaxis; thus if chemotaxis were inhibited inflammation could be reduced. In addition, cell motility is vital for metastatic disease and inhibitors of PI3K may also serve as therapeutic anticancer agents.

Phagocytosis and superoxide production are closely coupled in neutrophils. Current evidence demonstrates that PI3K play vital roles in both of these processes. Further studies should focus on the interaction of various classes of PI3K and the adapter molecules they recruit. Clearly, these are two fundamental aspects of cell biology and inflammation; a better understanding of the role of PI3K in these process could lead to novel therapeutic agents.

References

Bourne H, Weiner O (2002) Cell polarity: A chemical compass. Nature 419:21

Bravo J, Karathanassis D, Pacold C, Pacold M, Ellson C, Anderson K, Butler P, Lavenir I, Perisic O, Hawkins P, Stephens L, Williams R (2001). The crystal structure of the PX domain from p40(phox) bound to phosphatidylinositol 3-phosphate. Mol. Cell. 8:829–839

Bruyninckx W, Comerford K, Lawrence D, Colgan SP (2001). Phosphoinositide 3-kinase modulation of beta(3)-integrin represents an endogenous "braking" mechanism during neutrophil transmatrix migration. Blood 97:3251–3258

Cadwallader K, Condliffe A, McGregor A, Walker T, White J, Stephens L, Chilvers E (2002). Regulation of phosphatidylinositol 3-kinase activity and phosphatidylinositol 3,4,5-trisphosphate accumulation by neutrophil priming agents. J. Immunol. 169:3336–3344

Condliffe A, Hawkins P, Stephens L, Haslett C, Chilvers E (1998). Priming of human neutrophil superoxide generation by tumor necrosis factor-alpha is signaled by enhanced phosphatidylinositol 3,4,5-trisphosphate but not inositol 1,4,5-trisphosphate accumulation. FEBS Lett. 439:147–151

Corvera S (2001). Phosphatidylinositol 3-kinase and the control of endosome dynamics: new players defined by structural motifs. Traffic 2:859–866

Dekker, L Segal A (2000) Perspectives: signal transduction. Signals to move cells. Science 287:982–985

Dharmawardhane S, Brownson D, Lennartz M, Bokoch G (1999) Localization of p21-activated kinase 1 (PAK1) to pseudopodia, membrane ruffles, and phagocytic cups in activated human neutrophils. J. Leukoc. Biol. 3:521–527

Detmers P, Thieblemont N, Vasselon T, Pironkova R, Miller D, Wright S (1996) Potential role of membrane internalization and vesicle fusion in adhesion of neutrophils in response to lipopolysaccharide and TNF. J. Immunol. 157:5589–5596

Ellson C, Anderson K, Morgan G, Chilvers E, Lipp P, Stephens L, Hawkins P (2001a) Phosphatidylinositol 3-phosphate is generated in phagosomal membranes. Curr. Biol. 11:201–213

Ellson C, Gobert-Gosse S, Anderson K, Davidson K, Erdjument-Bromage P, Tempst P, Thuring J, Cooper M, Lim Z, Holmes A, (2001b). PtdIns(3)P regulates the neutrophil oxidase complex by binding to the PX domain of p40phox. Nature Cell Biol 3:679–692

Fruman D, Meyers R, Cantey L (1998). Phosphoinositide kinases. Annu. Rev. Biochem. 67:481–507

Fruman D, Cantley L (2002). Phosphoinositide 3-kinase in immunological systems. Sem. Immunol. 14:7–18

Funamoto S, Meili R, Lee S, Parry L, Firtel R (2002). Spatial and temporal regulation of 3-phosphoinositides by PI3-kinase and PTEN mediates chemotaxis. Cell 109:611–623

Funamoto S, Milan K, Meili R, Firtel R (2001). Role of phosphatidylinositol 3' kinase and a downstream pleckstrin homology domain-containing protein in controlling chemotaxis in *Dictyostelium*. J. Cell Biol. 153:795–810

Gerszten R, Friedrich E, Matsui T, Hung R, Li L, Force T, Rosenzweig A (2001). Role of phosphoinositide 3-kinase in monocyte recruitment under flow conditions. J Biol Chem. 276:26846–26851

Hannigan M, Zhan L, Ai Y, Wu D, Huang C (2002). Neutrophils lacking phosphoinositide 3-kinase gamma show loss of directionality during N-formyl-met-leu-phe-induced chemotaxis. Proc. Natl. Acad. Sci USA 99:3603–3608

Hiroaki H, Ago T, Ito T, Sumimoto H, Kohda D (2001). Solution structure of the PX domain, a target of the SH3 domain. 287:733–738

Hirsh E, Katanaev V, Garlanda C, Azzonilo O, Pirola L, Silengo L, Sozzani S, Mantovani A, Altruda F, Wymann M (2000). Central role for G protein-coupled phosphoinositide 3-kinase gamma in inflammation. Science 287:1049–1053

Iijima M, Huang Y, Deverotes P (2002). Temporal and spatial regulation of chemotaxis. Dev. Cell. 3:469–478

Kanai F, Liu H, Field S, Akbary H, Matsuo T, Brown G, Cantley L, Yaffe M (2001). The PX domains of p47phox and p40phox bind to lipid products of PI(3)K. Nature Cell Biol. 3:675–678

Karathanssis D, Stabelin R, Bravo J, Perisic O, Pcold C, Cho W, Williams R (2002). Binding of the PX domain of p47(phox) to phosphatidylinositol 3,4-bisphosphate and phosphatidic acid is masked by an intramolecular interaction. EMBO J. 21:5057–5068

Krugman S, Anderson K, Ridley S, Risso N, McGregor A, Coadwell J, Davidson K, Eguinoa H, Ellson D, Lipp P (2002). Identification of ARAP3, a novel PI3K effector regulating both Arf and Rho GTPases, by selective capture on phosphoinositide affinity matrices. Mol. Cell 9:95–113

Krystal G (2000). Lipid phosphatases in the immune system. Semin Immunol 12:397–403

Li Z, Jiang H, Xie W, Zuchuan Z, Smrka A, Wu D (2000). Roles for PLC-β2 and β3 and PI3Kγ in chemoattractant-mediated signal transduction. Science 287:1046–1049

Majeed M, Caveggion E, Lowell C, Berton G (2001). Role of Src kinases and Syk in Fcgamma receptor-mediated phagocytosis and phago-lysosome fusion. J. Leuk. Biol. 70:801–811

Ninomiya N, Hazeki K, Fukui Y, Seya T, Okada T, Hazeki O, Ui M (1994). Involvement of phosphatidylinositol 3-kinase in Fc gamma receptor signaling. J. Biol. Chem. 269:22732–22737

Noack D, Rae J, Cross A, Ellis B, Newburger P, Curnutte J, Heyworth P (2001). Autosomal recessive chronic granulomatous disease caused by defects in NCF-1, the gene encoding phagocyte p47-phox: mutations not arising in the NCF-1 pseudogenes. Blood 97:305–311

Perskvist N, Roberg K, Kulyte A, Stendahl O (1996). Rab5a GTPase regulates fusion between pathogen-containing phagosomes and cytoplasmic organelles in human neutrophils. J. Immunol. 157:5589–5596

Rappel W, Thomas P, Levine H, Loomis W (2002). Establishing direction during chemotaxis in eukaryotic cells. Biophys. J. 83:1361–1367

Rickert P, Weiner O, Wang F, Bourne H, Servant G (2000). Leukocytes navigate by compass: roles of PI3K gamma and its lipid products. Trends Cell Biol. 10:466–473

Roos D, Law S (2001). Hematologically important mutations: Leukocyte adhesion deficiency. Blood Cells Mol. Dis. 27:1000–1004

Rupper A, Grove B, Cardelli J (2001). Rab7 regulates phagosome maturation in *Dictyostelium*. J. Cell Sci. 114:2449–2460

Sasaki T, Irie-Sasaki J, Jones R, Oliveira-dos-Santos A, Stanford W, Bolon B, Wakeham A, Itie A, Bouchard D, Kozieradzki I, Joza N, Mak T, Ohashi P, Suzuki A, Penniger J (2000) Function of PI3Kγ in thymocyte development, T cell activation, and neutrophil migration. Science 287:1040–1046

Sato T, Overduin M, Emr S (2001). Location, location, location: Membrane targeting directed by PX domains. Science 294:1881–1885

Servant G, Weiner O, Herzmark P, Palla T, Sedat J, Bourne H (2000). Dynamics of a chemoattractant receptor in living neutrophils during chemotaxis. Science 287:1037–1040

Song X, Xu W, Zhang A, Huang G, Liang X, Virasius J, Czech M, Zhou G (2001). Phox homology domains specifically bind phosphatidylinositol phosphates. Biochemistry 40:8940–8944

Stephens L, Ellson C, Hawkins P (2002). Roles of PI3Ks in leukocyte chemotaxis and phagocytosis. Curr. Opin. Cell Biol. 14:203–214

Vieira O, Botelho R, Rameh L, Brachmann S, Matsuo T, Davidson H, Schreiber A, Backer J, Cantley L, Grinstein S (2001). Distinct roles of class I and class III phosphatidylinositol 3-kinase in phagosome formation and maturation. J. Cell Biol. 155:19–13

Vignais P (2002). The superoxide-generating NADPH oxidase: Structural aspects and activation mechanism. Cell Mol. Life Sci. 59:1428–1459

Weiner O, Neilson P, Prestwich G, Kirschner M, Cantley L, Bourne H (2002). A PtdInsP(3)- and Rho GTPase-mediated positive feedback loop regulates neutrophil polarity. Nat. Cell Biol. 4:509–513

Wu D, Huang C, Jiang H (2000). Roles of phospholipid signaling in chemoattractant-induced responses. J. Cell Sci. 113:2935–2940

Wymann M, Pirola L (1998). Structure and function of phosphoinositide 3-kinases. Biochem. Biophys. Acta 1436:127–150

Yaffe M (2002). The p47phox domain: two heads are better than one! Structure 10:1288–1290

Yasui K, Komiyama A (2001). Roles of phosphatidylinositol 3-kinase and phospholipase D in temporal activation of superoxide production in FMLP-stimulated human neutrophils. Cell Biochem. Funct. 19:43–50

Zhou K, Pandol S, Bokoch G, Traynor-Kaplan A (1998). Disruption of *Dictyostelium* PI3K genes reduces [32P]phosphatidylinositol 3,4 bisphosphate and [^{32}P]phosphatidylinositol trisphosphate levels, alters F-actin distribution and impairs pinocytosis. J Cell Sci 111:283–294

Nuclear Phosphoinositides and Their Functions

G. Hammond · C. L. Thomas · G. Schiavo

Molecular NeuroPathoBiology Laboratory, Lincoln's Inn Fields Laboratories, Cancer Research UK London Research Institute, 44 Lincoln's Inn Fields, London, WC2A 3PX, UK
E-mail: Giampietro.Schiavo@cancer.org.uk

1	Introduction	178
2	Structural and Functional Organisation of the Nucleus	179
3	Synthesis and Subnuclear Localisation of Phosphoinositides	183
3.1	PtdIns(4,5)P_2 and Its Synthetic Pathway	183
3.2	Phospholipase C and Diacylglycerol	186
3.3	PtdIns(3,4,5)P_3	187
3.4	PtdIns(3)P	188
4	Nuclear Roles of Phosphoinositides and Inositol Polyphosphates	189
4.1	Chromatin Remodelling and Control of Gene Expression	189
4.2	PtdIns(4,5)P_2 and Pre-mRNA Processing	193
4.3	mRNA Export	194
4.4	Cell Growth and Control of DNA Synthesis	196
4.5	Mitotic Regulation	197
4.6	Modulation of DNA Repair	198
5	Concluding Remarks	199
References		200

Abstract Phosphoinositides are minor components of biological membranes, which have emerged as essential regulators of a variety of cellular processes, both on the plasma membrane and on several intracellular organelles. The versatility of these lipids stems from their ability to function either as substrates for the generation of second messengers, as membrane-anchoring sites for cytosolic proteins or as regulators of the actin cytoskeleton. Despite a vast literature demonstrating the presence of phosphoinositides in the nucleus, only recently has the function(s) of the nuclear pool of these lipids and their soluble analogues, inositol polyphosphates, started to emerge. These compounds have been shown to serve as essential co-factors for several nuclear processes, including

DNA repair, transcription regulation and RNA dynamics. In this light, phosphoinositides and inositol polyphosphates might represent high turnover activity switches for nuclear complexes responsible for these processes. The regulation of these large machineries would be linked to the phosphorylation state of the inositol ring and limited temporally and spatially based on the synthesis and degradation of these molecules.

Abbreviations *DAG* Diacylglycerol · *IGC* Interchromatin granule cluster · *Ins* Inositol · *PC* Phosphatidylcholine · *PI* Phosphoinositides · *PIPK* PtdInsP kinase · *PLC* Phospholipase C · *PKC* Protein kinase C · *PtdIns* Phosphatidylinositol · *RNA Pol II* RNA polymerase II

1
Introduction

Although a relatively minor component of cellular membranes, phosphatidylinositol (PtdIns) has entered the limelight in cell biology over the last 20 years. This is because of the ability of its headgroup, inositol, to be phosphorylated at the D-3, -4 and -5 positions, yielding a family of seven distinct signalling molecules identified to date (Cockcroft 2000; Osborne et al. 2001a; Toker 2002).

The significance of phosphorylated PtdIns was first recognised with the discovery that the bisphosphorylated lipid phosphatidylinositol (4,5) bisphosphate [PtdIns(4,5)P_2] was hydrolysed at the plasma membrane to form the second messengers diacylglycerol (DAG) and inositol (1,4,5) trisphosphate (InsP$_3$) (Berridge 1993). However, in more recent years, PtdIns(4,5)P_2 itself has been shown to regulate a daunting number of cellular processes, including assembly and regulation of the actin cytoskeleton, exocytosis, endocytosis and intracellular trafficking to name but a few (Takenawa and Itoh 2001; Czech 2002; Toker 2002). PtdIns(4,5)P_2 appears to be localised in distinct subdomains of the plasma membrane, as well as in internal membranes such as the Golgi stacks and endoplasmic reticulum. Furthermore, its bisphosphorylated isomers PtdIns(3,4)P_2 and PtdIns(3,5)P_2 have recently been shown to have distinct biological functions. PtdIns(3,5)P_2 seems to localise at the multivesicular body (or vacuole in yeast) and to regulate vesicular traffic to this organelle (Cooke 2002). In contrast, PtdIns(3,4)P_2 is thought to be involved in anion superoxide production after phagocytosis in neutrophils (Cullen et al. 2001).

Phosphatidylinositol (3,4,5) trisphosphate [PtdIns(3,4,5)P$_3$], like PtdIns(4,5)P$_2$ regulates a plethora of cellular processes, which include actin dynamics and vesicular trafficking, as well as cell proliferation and survival (Toker 2002). PtdIns(3)P has also been shown to control several physiologic functions, including endosomal trafficking, ion channel regulation and cell survival (Gillooly et al. 2000; Cullen et al. 2001; Osborne et al. 2001a). In contrast, relatively little is known about the functions of the other monophosphorylated isomers, PtdIns(4)P and PtdIns(5)P, other than as precursors for the generation of PtdIns(4,5)P$_2$, although PtdIns(4)P-interacting proteins have recently been discovered (Czech 2002).

The soluble headgroup InsP$_3$ liberated from PtdIns(4,5)P$_2$ is also the precursor to a large family of inositol polyphosphates. These are now emerging as key players in a number of aspects of cellular physiology, including intracellular calcium homeostasis, ion channel physiology and membrane dynamics (reviewed in Irvine and Schell 2001).

Thus it is apparent that phosphoinositides (PI) play a key role in cytosolic physiology. However, despite a vast literature demonstrating the presence of PI in the nucleus, which are produced independently of the cytosolic cycle, relatively little is known about the functions of these molecules in this organelle (Irvine 2002). In the first part of this review, we describe the broad structure/function of the nucleus. We then annotate this map with the intranuclear locations of PI and the enzymes involved in their synthesis and degradation. In the last section, we discuss the various emerging roles for PI and inositol polyphosphates in nuclear function.

2
Structural and Functional Organisation of the Nucleus

Eukaryotic nuclei are not homogeneous but contain a variety of structures known as nuclear bodies, the majority of which are involved in the synthesis, processing and modification of RNA (Fig. 1A). Nuclear bodies lack the membrane boundaries that characterise cytoplasmic organelles, such as mitochondria and trans Golgi network, and are highly dynamic. The intrinsic flexibility and the appearance of nuclear bodies are dictated by the way in which nuclear RNA processing is organised in the nucleus. The transient association of transcription factors and RNA-processing complexes causes their accumulation on sites where high activity

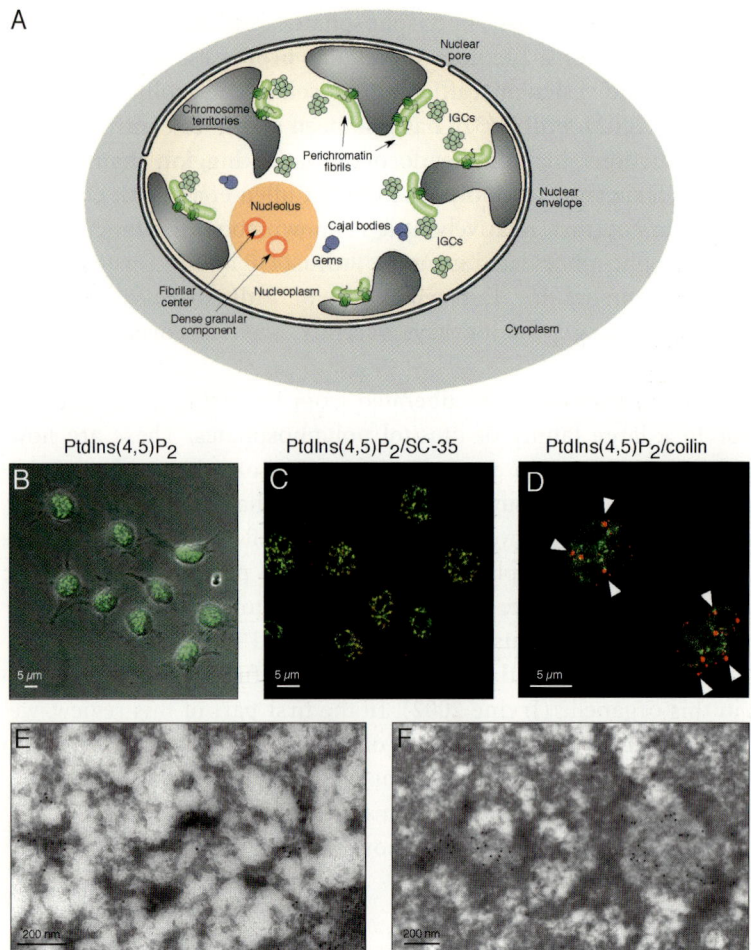

Fig. 1 A Scheme of the nucleus and its functional compartments. The chromosomes occupy discrete territories, with active sites of transcription located at their periphery and constituting the perichromatin fibrils. These functional transcription factories contain the transcriptional and pre-mRNA processing machinery, which also accumulates at the interchromatin granule clusters (IGCs). The maturation of small nuclear ribonucleoproteins (snRNPs) and small nucleolar RNPs (snoRNPs), a process which involves their assembly, modification and sorting, occurs at the Cajal bodies and their associated gemini of coiled bodies (or gems). Synthesis of ribosomes occurs in the nucleolus. In particular, the transcription of ribosomal DNA is thought to occur at the fibrillar centres, which are surrounded by the dense fibrillar component, whereas processing of rRNA and assembly of ribosomal proteins occurs out from these regions into the nucleolus. The entire nucleus is surrounded by the

is required, such as transcription hotspots. Therefore, nuclear bodies are the products of RNA processing activities rather than their pre-requisite (Lewis and Tollervey 2000).

The tight coupling between transcription and the major steps in pre-mRNA processing (capping, splicing, 3′-end processing and polyadenylation) are ensured by the recruitment of the processing complexes to the largest subunit of RNA polymerase II (RNA Pol II). The carboxy-terminal domain (CTD) of RNA Pol II acts as a major player in this process. The two major forms of RNA Pol II differ in the extent of CTD phosphorylation and are implicated in distinct phases of the transcriptosome activity. RNA Pol II containing a low phosphorylated CTD (RNA Pol IIa) is associated with transcription initiation, whereas the hyperphosphorylated form (RNA Pol IIo) is implicated in transcription elongation. The recruitment of pre-mRNA processing complexes on the CTD represents an efficient strategy for delivering these factors to the nascent pre-mRNA as it emerges from the polymerase, allowing spatial and temporal coupling of the process.

In interphase nuclei, chromosomes occupy discrete areas of the nucleus, termed chromosome territories. Areas of active transcription are normally distributed on the surface of these territories and account for 2,000–3,000 independent sites, which constitute the perichromatin fibrils (or fibres) (Fig. 1A). These sites are therefore functional "transcription factories" (Iborra et al. 2001) and contain high levels of essential splicing factors, such as SR proteins and spliceosomal small nuclear RNAs (snRNAs).

Several pre-mRNA processing factors are also accumulated in interchromatin granule clusters (IGCs or speckles) (Fig. 1A). These structures are better morphologically defined and less abundant than

◀─────────────────────────────────────

nuclear envelope, which is traversed by nuclear pore complexes, which regulate mRNA export and protein exchange with the cytosol. **B–F** Sub-nuclear localisation of PtdIns(4,5)P_2 using a specific monoclonal antibody 2C11 (*green*). PtdIns(4,5)P_2 is distributed in a speckled pattern in the nuclei of detergent extracted HeLa cells (**B**) and co-localises with the IGC marker SC-35 (*red*; **C**) but not with the Cajal bodies marker p80/coilin (*red, white arrowheads*; **D**). These PtdIns(4,5)P_2-enriched structures are decorated by 2C11 followed by 10-nm gold-conjugated secondary antibody, and appear electron dense in EM (**E**). PtdIns(4,5)P_2 is also present in the fibrillar centres and portions of the dense fibrillar component of nucleoli (**F**) perichromatin fibrils (20–50 IGCs/nucleus). Their appearance alters on blockade of transcription, which leads to the merging of IGCs in larger structures. Several lines of experimental evidence suggest that IGCs represent intermediate stations where complexes required for pre-mRNA processing are put together just before entering the site of active transcription. In the nuclear assembly line, IGCs are therefore nodal points in which essential factors converge and are fine-tuned before being sorted to neighbouring perichromatin fibrils.

The same paradigm seen for pre-mRNA synthesis and processing applies to the synthesis of ribosomes, a daunting task which is performed by the nucleolus (Lewis and Tollervey 2000). As seen for other nuclear bodies, the nucleolus self-assembles in association with ribosomal rDNA and concentrates all the components needed for ribosome synthesis, including RNA polymerase I. Formation of the nucleolus requires ongoing rRNA transcription, and its re-assembly after mitosis coincides with the start of pre-rRNA synthesis. However, several components of the nucleolus have the ability to associate independently of active transcription, highlighting the self-assembly property of this intranuclear organelle. Despite its static appearance in interphase cells, the nucleolus is very dynamic and pre-mRNA processing factors undergo a very rapid exchange with the surrounding nucleoplasm (Phair and Misteli 2000).

Cajal bodies (CB, also called coiled bodies) and their associated gems (gemini of coiled bodies) have been linked to the maturation of small ribonucleoproteins (snRNPs) and small nucleolar RNPs (snoRNPs) (Fig. 1A, D). Despite their highly structured appearance as revealed by electron microscopy, CB are also highly dynamic, their components exchanging rapidly with the surrounding nucleoplasm (Boudonck et al. 1999; Platani et al. 2002). This high transit rate prompted the hypothesis that, in addition to their role in the assembly and modification of snRNP and snoRNPs, these structures act as sorting stations for different classes of RNPs. To support their function as RNP factories, CBs tend to be located in close proximity to areas of chromatin coding for snRNP and snoRNPs (Lewis and Tollervey 2000).

3
Synthesis and Subnuclear Localisation of Phosphoinositides

3.1
PtdIns(4,5)P$_2$ and Its Synthetic Pathway

Perhaps the most fundamental discrepancy between the nuclear and cytosolic PI cycles is the fact that nuclear PI are located mainly outside of membrane bilayers. The first evidence for this came from the observation that Friend cell nuclei stripped of their envelopes contained PtdInsP and PtdIns(4,5)P$_2$ (Cocco et al. 1987). More recently, experiments utilising specific probes for PtdIns(4,5)P$_2$ have shed light on the specific localisation of this lipid in nuclei. Using an on-section labelling technique with the PtdIns(4,5)P$_2$-specific PH domain from PLCδ1, Watt and colleagues were able to show that the nucleus was the second most labelled structure after the plasma membrane within astrocytoma and A431 cells. However, relatively little PtdIns(4,5)P$_2$ was detected on the nuclear envelope, with the label being concentrated instead on electron-dense intranuclear structures (Watt et al. 2002) (Fig. 1E). Studies utilising monoclonal antibodies directed against PtdIns(4,5)P$_2$ reveal a punctate, "speckled" distribution by immunofluorescence (Mazzotti et al. 1995; Boronenkov et al. 1998; Osborne et al. 2001b) (Fig. 1B); this co-localised with components of the IGCs, including the splicing factor SC-35 (Fig. 1C) and RNA Pol IIo (Osborne et al. 2001b). PtdIns(4,5)P$_2$ was also found to associate with the fibrillar centre and the dense fibrillar component of the nucleolus (Fig. 1F) (Osborne et al. 2001b). Interestingly, a similar localisation was recently reported for PtdIns(3)P, suggesting that the nucleolus is a primary centre of PI accumulation in the nucleus (Gillooly et al. 2000).

The distribution of the nuclear PtdIns(4,5)P$_2$ is cell cycle dependent. On nuclear membrane disassembly, PtdIns(4,5)P$_2$ shifts to the cytoplasm and is excluded from the area occupied by the DNA (Osborne et al. 2001b). These PtdIns(4,5)P$_2$-containing structures remain distinct from other cytoplasmic organelles undergoing mitotic fragmentation, such as the Golgi apparatus, during all stages of the cell cycle. In late telophase, PtdIns(4,5)P$_2$ accumulates in discrete structures, which are positive for RNA Pol IIo and SC-35. These structures, which resemble mitotic interchromatin granules (Leser et al. 1989; Spector et al. 1991), remain peripheral despite the reformation of the daughter cell nuclei. At present, the precise mechanism leading to their re-entry in the nucleus is unclear. In particular, it is unknown whether these lipid complexes are transferred through the nuclear pore on their disassembly or retrieved to the nucleoplasm via fenestrations still present in the nuclear envelope (Nakielny and Dreyfuss 1999).

The nuclear envelope has been found to project long invaginations that penetrate deeply inside the nucleus (Fricker et al. 1997). To rule out the possibility that the intranuclear PtdIns(4,5)P_2 was located at these deep invaginations, Boronenkov and colleagues stained these structures with concanavalin-A, and observed no co-localisation. (Boronenkov et al. 1998). What is the structural context of PtdIns(4,5)P_2 and other intranuclear lipids if they are not present in membrane bilayers? Presumably, some form of proteolipid complexes, which in the case of PtdIns(4,5)P_2 include nucleic acids (Osborne et al. 2001b), are responsible for the stabilisation of PIs under these conditions and might explain their resistance to detergent extraction. In this regard, PtdIns(4,5)P_2 has been shown to interact strongly with proteins in the absence of a membrane bilayer (Fukami et al. 1992; Fukami et al. 1996; Liu et al. 1996; Cockcroft 1998).

In support of this extra-membranous localisation of PtdIns(4,5)P_2, several lines of evidence indicate that the kinases responsible for synthesis of this lipid are present in the nuclear matrix. PtdIns(4,5)P_2 is now known to be synthesised via two routes. The first is by phosphorylation of PtdIns(4)P at the D-5 position by PtdIns(4)P 5-kinases, the so-called type I PtdInsP kinases (PIPK I); the second is by phosphorylation of PtdIns(5)P at the D-4 position by the type II PtdInsP kinases (PIPK II). PtdIns(4)P is produced by the family of PtdIns 4-kinases, although the route of synthesis of PtdIns(5)P is as yet undefined. Rat liver nuclei stripped of their envelopes with detergent retain around 40% of the mass of PtdIns(4,5)P_2 found in intact nuclei (Vann et al. 1997). When exogenous PtdIns and PtdIns(4)P were added to these stripped nuclei, around 70% of the PtdIns 4-kinase and 90% of the PtdIns(4)P 5-kinase activities were retained, indicating their resistance to detergent extraction. Immunofluorescence of endogenous and over-expressed PIPK Iα and PIPK IIα in detergent-extracted NRK and 2RA cells revealed a diffuse nuclear staining enriched at snRNA and PtdIns(4,5)P_2-containing speckles (Boronenkov et al. 1998). Furthermore, transient transfection of HeLa cells reveals that GFP-tagged PIPK IIβ is localised in the nucleus, although the isoform α is excluded and instead has a cytosolic localisation

(Ciruela et al. 2000). This distribution was found to be driven by alpha helix 7, which is absent in the α isoform (Ciruela et al. 2000). How then could PIPK IIα enter nuclei without this localising helix? A plausible explanation foresees PIPK IIα obtaining a "piggy-back" on a PIPK I, as shown previously for the translocation of PIPK IIα to the plasma membrane (Hinchliffe et al. 2002).

The substrates for PIPK I and II have been identified in the nuclei of MEL cells. These are regulated by different mechanisms, because although the mass of both PtdIns(4)P and PtdIns(5)P are seen to decrease at S-phase, the reduction in mass of the latter is around an order of magnitude greater (Clarke et al. 2001). However, in *Dictyostelium* there is evidence for at least some redundancy between these two substrates. Early development in *Dictyostelium* appears to be regulated by an enzyme with PIPK activity, designated PIPKinA (see below). This enzyme cannot easily be classified as a type I or type II PIPK; indeed, the mutant phenotype in *Dictyostelium* can be rescued by transfection of cells with PIPKinA where the catalytic domain has been replaced with either mammalian PIPK type Iβ or type IIβ domains (Guo et al. 2001).

Despite convincing evidence for an intranuclear localisation of PtdIns(4,5)P_2, its precursors and biosynthetic machinery, the topology of its synthesis remains unclear. Isolated Swiss 3T3 nuclei show PtdIns 4-kinase and PIPK activity, but if the peripheral and inner matrices are separated PtdIns kinase activity is found only in the periphery whereas PIPK activity is found within the inner matrix (Payrastre et al. 1992). Moreover, whereas PtdIns(4)P and PtdIns(4,5)P_2 are present in nuclei stripped of their envelopes, PtdIns is undetectable (Vann et al. 1997). Thus PtdInsP is apparently synthesised at sites distinct from the IGCs where PtdIns(4,5)P_2 is present and must be transported from one site to another. It will be particularly interesting in the future to elucidate the sub-nuclear distributions of the PtdIns 4-kinases, as well as PtdIns(4)P and PtdIns(5)P. It is interesting to note that the α isoform of phosphatidylinositol transfer protein (PITP) is enriched in the nuclear matrix (De Vries et al. 1996) and may be responsible for the transport of these lipids.

3.2
Phospholipase C and Diacylglycerol

The first physiologic function elucidated for PtdIns(4,5)P_2 was in the generation of the second messengers DAG and InsP$_3$ by members of the PtdIns-specific phospholipase C family (PI-PLC). Indeed several PI-PLC isoforms (β, γ and δ) have been identified in the nuclei of a variety of cell types (reviewed in D'Santos et al. 1998). Once generated, DAG can function in nuclei to activate various protein kinase C (PKC) isoforms (Martelli et al. 1999), whereas InsP$_3$ may function to regulate nuclear Ca^{2+}, or serve as a substrate for the generation of higher inositol polyphosphates (Irvine 2002), which have a number of nuclear functions. However, studies of the DAG species produced in the nuclei of MEL and Swiss 3T3 cell nuclei showed that the majority of DAG contained a fatty acid profile more consistent with phosphatidylcholine (PC) (D'Santos et al. 1999). Like PtdIns(4,5)P_2, PC is retained in nuclei after envelope removal with detergent (Maraldi et al. 1992; Vann et al. 1997). Generation of DAG from PC could occur via two routes: (1) a specific PC-PLC activity, so far not identified in eukaryotes. or (2) phospholipase D-mediated production of phosphatidic acid (PA), followed by dephosphorylation to yield DAG (D'Santos et al. 1998). However, the nuclei of MEL cells do have a PI-PLC activity, which appears to be cell cycle regulated and leads to the synthesis of PA via DAG phosphorylation (D'Santos et al. 1999).

Why do nuclei have two apparently independent routes for synthesis of DAG? We can envisage this being due to at least three non-mutually exclusive scenarios: (1) DAG generated by PI-PLC is rapidly phosphorylated to PA, causing a transient activation of PKC, whereas PC-mediated production of DAG generates a more stable pool which can lead to sustained activation of PKC. (2) DAG-mediated activation of PKC may depend on the fatty acid content of the lipid (Martelli et al. 2003). Consistent with this possibility is the observation that in HL-60 cells, IGF-I activates nuclear PLC-βI to generate PtdIns(4,5)P_2-derived DAG, which in turn activates PKC-β_{II}; on the other hand, DMSO-induced differentiation leads to PC-derived DAG, which activates PKC α (Neri et al. 2002). (3) PC-derived DAG may act as a nuclear-specific mechanism for PKC activation, bypassing the concomitant production of InsP$_3$ and its phosphorylated derivatives, which display a number of nuclear functions (see below).

3.3
PtdIns(3,4,5)P$_3$

Another second messenger derived from PtdIns(4,5)P$_2$ is PtdIns(3,4,5)P$_3$. Its production is catalysed primarily by the type I phosphatidylinositol 3-kinases (PI3K), which have been identified in nuclei (Zini et al. 1996; Bertagnolo et al. 1998; Calcerrada et al. 2002). Isolated rat liver nuclei show incorporation of ^{32}P into PtdIns(3,4,5)P$_3$ after incubation with [γ-^{32}P]-ATP, although this process is abolished by extraction of the envelope with detergent (Sindic et al. 2001). This suggests that PtdIns(3,4,5)P$_3$ may be generated in nuclei, but only at the level of the envelope. A lack of incorporation of ^{32}P into D-3 PI was shown in isolated smooth muscle cell nuclei in resting conditions (Bacqueville et al. 2001). However, production of PtdIns(3,4,5)P$_3$ is observed when these nuclei are incubated with GTPγS; this can be abolished by incubation with the PI3K inhibitors wortmannin and LY294002, as well as by pertussis toxin, implicating G proteins in the regulation of the nuclear PI3K. The γ isoform of the type I PI3K (PI3K Iγ) is expressed in nuclei of these cells (Bacqueville et al. 2001), which is indeed known to be regulated by heterotrimeric G proteins (Vanhaesebroeck et al. 2001). In support of this, Metjian et al. reported translocation of PI3K Iγ into the nuclei of hepatoma cells in response to serum, which is blocked by pertussis toxin and can be rescued by over-expressing the heterotrimeric G protein subunit G$_{\beta\gamma}$ (Metjian et al. 1999).

PI3K Iγ is apparently not the only isoform of PI3K in nuclei. Lu and colleagues recently showed expression of the p85 regulatory subunit of the type IA PI3K in rat liver nuclei by immuno-electron microscopy (Lu et al. 1998). Some members of this class of PI3K are also regulated by G proteins (Vanhaesebroeck et al. 2001). Interestingly, Ye et al. identified a novel nuclear GTPase, PIKE, which was able to activate the type IA PI3K in HEK293 cells (Ye et al. 2000).

Further evidence for a sub-nuclear localisation of PtdIns(3,4,5)P$_3$ comes from studies showing the distribution of the phosphatase, PTEN, that degrades PtdIns(3,4,5)P$_3$ back to PtdIns(4,5)P$_2$. PTEN has been shown to be expressed in the nuclei of thyroid epithelial cells (Gimm et al. 2000) and at speckles in NGF-differentiated PC12 cells (Lachyankar et al. 2000). The nuclear pool of PtdIns(3,4,5)P$_3$, is likely to be synthesised in a regulated manner, rather than being present constitutively as is observed for PtdIns(4,5)P$_2$. Although specific functions for this nuclear pool of PtdIns(3,4,5)P$_3$ have not been defined, it may contribute to the PI-dependent modulation of Akt/PKB (Brazil and Hemmings 2001), which has been shown to translocate to the nucleus (Borgatti et al. 2000).

3.4
PtdIns(3)P

Recently, evidence that the monophosphorylated PI, PtdIns(3)P, is produced in nuclei has been presented. Embryonic carrot nuclei were able to produce PtdIns(3)P even after detergent extraction (Bunney et al. 2000), and a PtdIns(3)P-specific probe derived from a tandem fusion of two FYVE domains revealed the presence of PtdIns(3)P in the nucleolus of mammalian cells, specifically at the dense fibrillar component (Gillooly et al. 2000). PtdIns(3)P is synthesised by the class II and III PI3Ks (Vanhaesebroeck et al. 2001). In fact, the type III PI3K from soybean (SPI3K-5p) was found to have an intranuclear localisation and to accumulate at sites of active transcription (Bunney et al. 2000). Moreover, the type II isoform PI3K C2α has been shown to be localised at speckles in HeLa cell nuclei and its distribution overlaps with the snRNA marker Sm even after inhibition of transcription with α-amanitin (Didichenko and Thelen 2001). This staining was resistant to detergent or salt extraction, as seen in the case of PIPK (Boronenkov et al. 1998). Furthermore, HL-60 cells differentiated with all-*trans* retinoic acid showed an increase in PtdIns(3)P and PtdIns(3,4,5)P$_3$ synthesis, as well as an increase of nuclear PI3K C2β activity (Visnjic et al. 2002). Substituting calcium for magnesium in the kinase assay inhibited PtdIns(3,4,5)P$_3$ but not PtdIns(3)P production, implicating type I PI3K in the production of PtdIns(3,4,5)P$_3$ and PI3K C2β in production of PtdIns(3)P. Note that in contrast to Didichenko and Thelen's study, this PI3K activity appeared enriched on the nuclear envelope. PI3K C2β activity was also shown in regenerating rat liver cell nuclei and correlated with the onset of PtdIns(3)P production (Sindic et al. 2001). Further hints as to a function for nuclear PtdIns(3)P come from analysis of the *Arabidopsis* protein database, which reveals three proteins with putative nuclear functions containing FYVE domains. However, two of these FYVE domains are variant and may bind other targets such as PtdIns(5)P (Drobak and Heras 2002). Thus it appears that PtdIns(3)P has multiple nuclear localisations, in the matrix, nucleolus and possibly on the nuclear envelope.

4
Nuclear Roles of Phosphoinositides and Inositol Polyphosphates

In recent years, the number of nuclear pathways that have been demonstrated to involve PI and inositol polyphosphates has increased massively. Many aspects of the regulation of nucleic acid synthesis, processing and repair have been shown by genetic and/or biochemical studies to entwine with the PI pathway (Fig. 2), although in several instances the nature of the inositol derivative and its precise role are not completely clear.

4.1
Chromatin Remodelling and Control of Gene Expression

PtdIns(4,5)P_2 shows interactions with several nuclear proteins involved in distinct functions. In particular, PtdIns(4,5)P_2 binds to histones H1 and H3 (Yu et al. 1998), two very abundant components of eukaryotic chromatin (Horn and Peterson 2002). In eukaryotic cells, DNA bound to histones is packaged into chromatin, which forms a repressive structure limiting the access of the transcription machinery to DNA. The building block of chromatin is the nucleosome, which contains 147 base pairs of DNA wrapped around a core histone octamer containing two copies each of the histones H2A, H2B, H3 and H4 (Luger et al. 1997; Felsenfeld and Groudine 2003). In addition to the core histones, metazoan chromatin also contains linker histones (such as histone H1), which bind to nucleosomes and 20 base pairs of DNA. Linker histones provide a structured connection between nucleosomes and play a major role in maintaining higher-order chromatin structure. Core and linker histones share a basic structure, with a globular domain flanked by amino- and carboxy-terminal tail domains. The tail domains are believed to be important for post-translational modifications and, in the case of histone H1 and other linker histones, in chromatin folding (Horn and Peterson 2002). PtdIns(4,5)P_2 binds to the carboxy-terminal domain of histone H1, and this interaction is inhibited specifically on phosphorylation by PKC. Interestingly, PKC phosphorylation of histone H1 has been shown to occur on cell stimulation (Divecha et al. 1993; Hocevar et al. 1993). PtdIns(4,5)P_2 and to a lesser extent PtdIns(4)P and PtdIns(3,4,5)P_3, but not other PI nor InsP$_3$, counteract the histone H1-mediated repression of the basal transcription in a *Drosophila* transcription system. These results suggest that translocation of activated PKCs into the nucleus might

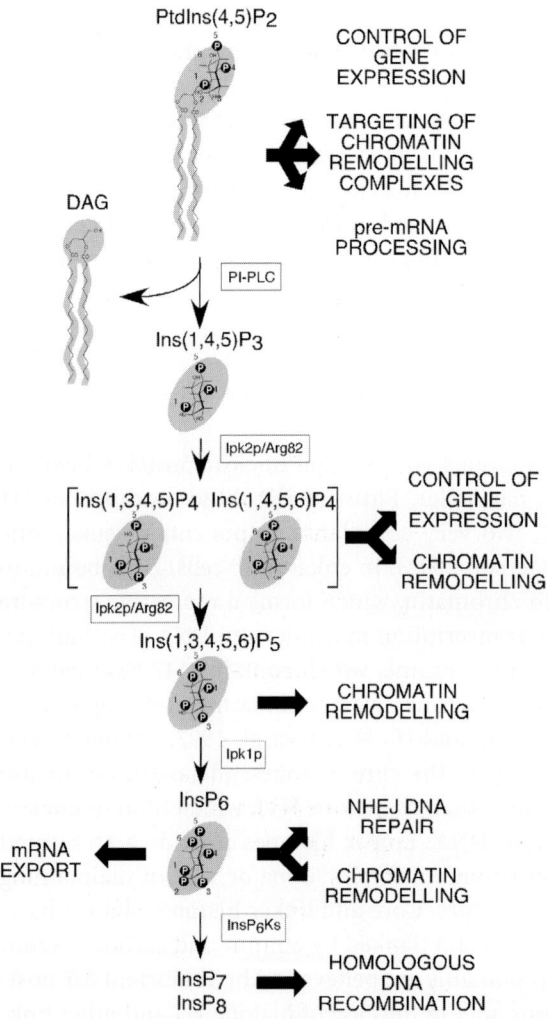

Fig. 2 Scheme of the synthetic pathway of inositol polyphosphates in yeast nuclei. Enzymes involved at each step of the pathway are *boxed*, and *black arrows* point to some of the identified functions for each molecule

lead to histone phosphorylation and release of PtdIns(4,5)P_2, which would be available for PLC processing. Conversely, synthesis of nuclear PtdIns(4,5)P_2 might alter the binding of histone H1 from DNA, relieving its inhibitory effect on transcription (Yu et al. 1998).

Evidence that enzymes involved in nuclear PtdIns(4,5)P_2 participate in the control of gene expression is also emerging. Guo and collaborators demonstrated that abrogating the activity of a novel PIPK in *Dictyostelium* inhibits developmental gene expression, causing a block at an early stage of development (Guo et al. 2001). This novel enzyme, termed PIPkinA, is enriched in the nucleus and has an atypical structure. Mutants null for PIPkinA do not aggregate in response to starvation, which is not due to an impaired chemotactic response to cAMP but rather to a low level of expression of its receptor (Guo et al. 2001). Early events in cAMP signalling in *Dictyostelium* include the activation of a PLCδ family member leading to the generation of DAG. The characterisation of PIPkinA is particularly important because this enzyme contains a selective DAG-binding domain, which does not interact with the phorbol ester derivative phorbol 12,13-dibutyrate (PDBu). Although additional work is still necessary to assess the full impact of this enzyme in the synthetic pathway of nuclear PI (e.g. the specificity of its PIPK activity has not been investigated yet), these findings suggest that this enzyme is a candidate effector for DAG in the nucleus, which regulates gene expression in early development.

A direct link between a nuclear PI cycle and control of gene expression has been recently provided by the finding that chromatin remodelling complexes show biochemical and genetic interactions with inositol polyphosphates and the inositol polyphosphate kinase Ipk2p. By altering the binding of histones with DNA, chromatin remodelling complexes promotes efficient transcription of eukaryotic genes. How these complexes are targeted to specific sites of chromatin and the molecular basis of their regulation are presently unknown. A genetic screen for *S. cerevisiae* mutants that were defective in induction of the phosphate-responsive gene PHO5 identified the nuclear inositol polyphosphate kinase ARG82/IPK2 as a target gene (Steger et al. 2003). Ipk2p is a catalytically flexible InsP$_3$-InsP$_4$ kinase, which phosphorylates InsP$_3$ in positions 3 and 6, generating Ins(1,3,4,5,6)P$_5$ as its end product (Shears 2000). In Arg82 mutant strains, remodelling of PHO5 promoter chromatin is impaired and the ATP-dependent chromatin-remodelling complexes SWI/SNF and INO80 are not efficiently recruited to phosphate-responsive promoters (Steger et al. 2003). With an independent biochemical approach, Shen and colleagues confirmed these data and found that inositol polyphosphates directly modulate several classes of chromatin remodelling complexes in metazoans in vitro and in vivo (Shen et al.

2003). In particular, $InsP_6$ and $InsP_5$, but not other inositol polyphosphates, inhibit the nucleosome mobilisation promoted by the ISWI-containing *Drosophila* NURF complex. This effect occurs by direct inhibition of its ATPase activity and is specific, because other unrelated ATPases are unaffected. The modulatory action of $InsP_6$ extends to the yeast ISW2, demonstrating that this property is conserved through evolution. The same inhibitory effect of $InsP_6$ was observed on another chromatin remodelling complex, INO80, at the level of both ATPase activity and nucleosome mobilisation. In sharp contrast, $InsP_6$ was completely ineffective on the yeast SWI/SNF complex, whereas both $Ins(1,4,5,6)P_4$ and $InsP_5$ strongly stimulated its chromatin remodelling activity. In addition, the ATPase activity of SWI/SNF is not affected by inositol polyphosphates, indicating that the mechanism whereby these compounds modulate the chromatin remodelling activity of the SWI/SNF complex is different from that seen for INO80 and NURF (Shen et al. 2003). Given that some genes show differential regulation by distinct chromatin remodelling complexes, cellular levels of inositol polyphosphates could fine-tune the balance between synergistic and antagonistic chromatin-remodelling activities (Fig. 2). Inositol polyphosphates may therefore represent essential players in the regulation of chromatin remodelling and gene expression.

The inositol-dependent regulation of chromatin remodelling is not limited to inositol polyphosphates but extends to PI. In particular, the SWI/SNF-like BAF chromatin remodelling complex binds to $PtdIns(4,5)P_2$-containing liposomes (Rando et al. 2002). The interaction of this complex to chromatin is also $PtdIns(4,5)P_2$ dependent (Zhao et al. 1998). $PtdIns(4,5)P_2$ modulates the binding of the SWI/SNF-like BAF complex to F-actin pointed ends and branchpoints. Furthermore, actin polymerisation is promoted in a $PtdIns(4,5)P_2$-dependent manner in the presence of the SWI/SNF-like BAF complex. Actin displays two distinct binding sites on the ATPase subunit of the complex, which are differentially regulated by $PtdIns(4,5)P_2$. Interestingly, several of the SWI/SNF-like complexes contain at least one actin-related protein (Arp). In addition to β-actin, the mammalian SWI/SNF-like BAF complex contains BAF53, which is closely related to Arp3, whereas Arp7 and Arp9 are functional components of the yeast SWI/SNF complex (Cairns et al. 1998). This property is not exclusive to the SWI/SNF-like complexes because the chromatin remodelling enzyme Ino80 has been also shown to contain Arps as well as actin.

Altogether these findings point to a strong connection between chromatin remodelling, the PI cycle and actin, renewing the debate on the functional roles of actin in the nucleus (Rando et al. 2000) and its relationships with the nuclear matrix (Pederson 2000). Although evidence supporting the presence of a nuclear F-actin pool has not been shown, actin could be present in the nucleus as short oligomers constituting a highly dynamic matrix (Rando et al. 2000). By enhancing the interaction of SWI/SNF-like BAF complex with actin, PtdIns(4,5)P_2 might function as a matrix localisation signal for these enzymes. In particular, PtdIns(4,5)P_2 might directly activate F-actin recruitment of the BAF complex by interacting with its ATPase subunit and therefore relieving the capping on endogenous BAF53 and β-actin. The uncapped complex would be competent for binding with F-actin via its bound endogenous Arps, in a process which might lead to force generation and to the resolution of high-order chromatin structures (Rando et al. 2002).

4.2
PtdIns(4,5)P_2 and Pre-mRNA Processing

By using a PtdIns(4,5)P_2-specific antibody (2C11) (Thomas et al. 1999), Osborne and colleagues found that in interphase mammalian cells PtdIns(4,5)P_2 assembles in nuclear complexes containing proteins, lipids and nucleic acid, which are resistant to detergent extraction (Osborne et al. 2001b). These electron-dense particles are reminiscent of IGCs and contain components of the transcriptional and pre-mRNA processing machinery, including RNA Pol IIo and SC-35 (Fig. 1C) (Boronenkov et al. 1998; Osborne et al. 2001b). During interphase, PtdIns(4,5)P_2-containing structures are unlikely to contain DNA because they are resistant to DNAse treatment. By contrast, incubation with RNAse completely abolishes PtdIns(4,5)P_2 staining, indicating that RNA is essential for the stability of the PtdIns(4,5)P_2-containing complexes (Osborne et al. 2001b). Given their close resemblance to IGCs and their intrinsic features, it is likely that these particles are involved in pre-mRNA processing. Immunodepletion experiments of nuclear extracts with anti-PtdIns(4,5)P_2 antibody indicate that PtdIns(4,5)P_2-containing structures are indeed essential for pre-mRNA splicing and that they are associated with the main pre-mRNA splicing activity in mammalian cells (Osborne et al. 2001b). However, PtdIns(4,5)P_2-associated complexes are not sufficient to sustain pre-mRNA splicing in vitro, suggesting that other nuclear factors are necessary for this process.

What is the function of PtdIns(4,5)P_2 in these complexes? The lack of stimulation of pre-mRNA splicing by PtdIns(4,5)P_2 strongly suggests that this lipid is not having a direct regulatory role on the activity of the spliceosome. Three alternative, but non-mutually exclusive, functions can be envisaged for this nuclear PtdIns(4,5)P_2 pool. The first foresees PtdIns(4,5)P_2 as a modulator of the distribution and mobility of the spliceosome. This might be achieved by the interaction of this PI with nuclear cytoskeletal proteins, such as actin and actin-binding proteins (Rando et al. 2000), and other PtdIns(4,5)P_2-binding factors, such as protein 4.1, which has been shown to functionally interact with the splicing apparatus (Lallena and Correas 1997). PtdIns(4,5)P_2 would therefore serve as a structural interface between the enzymatic core of the spliceosome, the sites of active transcription and nuclear sub-domains (Osborne et al. 2001b). Second, PtdIns(4,5)P_2 might be used as a substrate by nuclear PLCs as a source of localised pools of InsP$_3$ and DAG (Irvine 2002). Besides its function as second messenger, InsP$_3$ can be used for the production of inositol polyphosphates, which in turn act as co-factor/regulator of several nuclear processes (Fig. 2). Finally, PtdIns(4,5)P_2 could specifically regulate the activity of some spliceosomal components. PtdIns(4,5)P_2 modulates casein kinase Iα, which localises on IGCs. Casein kinase Iα was shown to phosphorylate members of the SR family of splicing factors (Gross et al. 1999), suggesting that PtdIns(4,5)P_2 might influence pre-mRNA processing via direct modulation of nuclear kinases.

4.3
mRNA Export

In eukaryotic cells, transcription/pre-mRNA processing and translation are both temporally and spatially segregated by the nuclear membrane. The *trait d'union* between the two processes is the nuclear pore complex, which allows communication between the nuclear environment and the cytosol with the exchange of ions, RNAs and proteins (Nakielny and Dreyfuss 1999). mRNA export from the nucleus includes a cascade of events, which ensures that only mature RNA is exported. RNA in complex with RNA-binding proteins is then targeted to the nuclear pore and translocated to the cytosol in an ATP-dependent process (Nakielny and

Dreyfuss 1999). Pre-mRNA processing and transport are highly coordinated events, because blockade of any of the maturation steps affects the export (Lee and Silver 1997).

A functional connection between mRNA export and PI dynamics in the nucleus has been shown by York and colleagues, who reported a genetic interaction between a Gle1p, a protein associated with the nuclear pore complex essential for mRNA export, and three genes involved in PI metabolism (York et al. 1999). Mutations in these genes block mRNA nuclear export. The first gene is PLC1, which encodes the yeast orthologue of the PtdIns-specific phospholipase C, whereas the second (GSL1) encodes for an $InsP_5$ 2-kinase named Ipk1p. This kinase converts $Ins(1,3,4,5,6)P_5$ to $InsP_6$ and is localised at the nuclear envelope and in the proximity of the nuclear pore. Inactivation of Plc1p and Ipk1p abolished $InsP_6$ synthesis, suggesting that this soluble inositol polyphosphate is essential for the nuclear export of mRNA (Fig. 2). Moreover, it suggests that, at least for nuclear export, $PtdIns(4,5)P_2$ is required as a substrate to ensure inositol polyphosphate synthesis (York et al. 1999). The third gene identified in this investigation is a component of the arginine-responsive ArgR-Mcm1 complex, which comprises four proteins, all required for proper transcriptional control (Saiardi et al. 1999; Odom et al. 2000; Saiardi et al. 2000). Whereas two of these factors are involved in arginine-specific transcription in S. cerevisiae, *Arg82*, the gene isolated in the screening, and *Mcm1* encode pleiotropic regulators (Shears 2000). *Arg82* mutants have defects in mating, sporulation and response to stress and nutrients. Arg82p is also known as Ipk2p (inositol polyphosphate kinase) (Odom et al. 2000) and inositol polyphosphate multikinase (ipmkp) (Saiardi et al. 1999). Ipk2 mutation causes the accumulation of $InsP_3$ and loss of intracellular $InsP_6$. Ipk2p is therefore the main $InsP_3$ kinase activity in yeast (Odom et al. 2000; Saiardi et al. 2000).

Ipk2p, but not its catalytic activity, is required for the assembly of the ArgR-Mcm1 complex. However, the Ipk2p-dependent production of $InsP_4$ is necessary in vivo for the transcriptional activation mediated by ArgR-Mcm1 (Odom et al. 2000). Is this a new function for $InsP_4$ in the nucleus? The experiments by Odom and colleagues support this conclusion and suggest that $InsP_4$ may influence transcriptional control (Fig. 2). The requirement for a kinase activity may empower the ArgR-Mcm1 complex to carry out transcription. Therefore, Ipk2p influences transcriptional responses and mRNA availability by three mechanisms:

(1) formation of ArgR-Mcm1 complex; (2) transcriptional activation in vivo via production of $InsP_4$ and (3) mRNA export by entering the pathway involving PLC1 and Ipk1p, which leads to the production of $InsP_6$ from $PtdIns(4,5)P_2$.

Given these results, to attribute a reduced rate of mRNA export to a decrease in the intracellular concentration of $InsP_6$ could represent an over-simplification, both from the point of view of the multiple functional interactions demonstrated for Ipk2p and the fact that $InsP_6$ does not represent a metabolic end-point in yeast (Saiardi et al. 2000). Furthermore, it is unclear whether these mechanisms are functional in higher eukaryotes (Shears 2000). However, orthologues of Ipk1p and Ipk2p are widely distributed in nature, including humans and plants (York et al. 1998; Chang et al. 2002; Stevenson-Paulik et al. 2002; Verbsky et al. 2002). More strikingly, the over-expression of the inositol phosphate phosphatase SopB, a virulence factor in *Salmonella*, in mammalian cells causes the depletion of inositol phosphates, including $InsP_5$ and $InsP_6$. This depletion was coincident with the accumulation of polyadenylated RNA in the nucleus, suggesting a defect in nuclear export. Localisation of SopB to the nucleus is strictly required for the mRNA transport defect to occur (Feng et al. 2001). These results provide evidence that inositol phosphate synthesis is also necessary for efficient mRNA export in mammalian cells.

4.4
Cell Growth and Control of DNA Synthesis

Nuclear PI levels, including those of $PtdIns(4)P$ and $PtdIns(4,5)P_2$, decrease strikingly during S-phase (York and Majerus 1994). This reflects the activation of a nuclear PLC, likely to be a member of the PLCβ family, whose activity increases during S-phase (Kuriki et al. 1992). In fact, PLCβ1 has been linked to mitogen-activated cell growth and its activity increases two- to threefold on IGF-I stimulation of Swiss 3T3 cells (Martelli et al. 1992; Manzoli et al. 1999). This causes the breakdown of nuclear PI and DAG release (Divecha et al. 1991), which in turn is required for IGF-I-dependent nuclear translocation of PKC (Martelli et al. 1991). Downregulation of PLCβ1 expression blocks DNA synthesis, whereas its overexpression increases the number of cells in S-phase after IGF-I treatment (Manzoli et al. 1997). Nuclear PLC β1 has been shown to be phosphorylated on IGF-I mediated stimulation of Swiss 3T3 cells;

cells over-expressing mutant PLC β1 that lacks this phosphorylation site do not activate nuclear PtdIns(4,5)P$_2$ hydrolysis, nor do cells enter S-phase (Xu et al. 2001). The same applies to the activation of osteosarcoma cells by interleukin 1α, which suggests that nuclear PLCβ1 activation and PtdIns(4,5)P$_2$ hydrolysis are very early events in the signalling cascade evoked by mitogens and cytokines. Another interesting candidate, possibly acting at later stages, is PLCδ4, which is dramatically up-regulated during G_1 to S-phase transition (Liu et al. 1996). Conversely, cell differentiation causes a block in DNA synthesis and a parallel increase in nuclear PtdIns(4)P and PtdIns(4,5)P$_2$ (Cocco et al. 1987).

DNA synthesis might also be regulated by nuclear inositol polyphosphates and by an inositol pathway involving the enzyme inositol polypolyphosphate 1-phosphatase (1-ptase). 1-ptase dephosphorylates Ins(1,4)P$_2$ and Ins(1,3,4)P$_2$ by removing the phosphate group in position 1 (York et al. 1998). This enzyme is concentrated in the nucleus and its activity is inhibited by lithium. Both substrates and products of 1-ptase are intermediates of the inositol polyphosphate cycle, but, with the notable exception of Ins(1,4)P$_2$, which regulates DNA polymerase α (Sylvia et al. 1988), their interacting proteins are presently unknown. Although the precise molecular mechanism of 1-ptase's cellular action is still unclear, its inhibitory effect on DNA synthesis suggests that entry in S-phase is accompanied by the production of inositol polyphosphates generated by the action of PLCs on PtdIns(4,5)P$_2$. The promoting activity of inositol polyphosphates on DNA synthesis is counteracted by 1-ptase and other putative nuclear inositol phosphatases, which may regulate S-phase via their hydrolytic activity.

4.5
Mitotic Regulation

Evidence for a role of the PI cycle in the regulation of the G_2/M phase transition has also been reported. In particular, nuclear levels of DAG rise to a peak coincident with the G_2/M transition before returning to basal levels during G_1. The increase in the nuclear concentration of DAG at this point in the cell cycle is caused by a PI-specific PLC activity. Importantly, inhibition of this PLC activity causes G_2 arrest (Sun et al. 1997). Neri and collaborators found that in mammalian cells, IGF-I induced PI-PLC-derived DAG production, translocation of PKCα into the nucleus and phosphorylation of histone H1 (Neri et al. 1998). Moreover, in mouse oocytes, PLC β1 accumulates in the nucleoplasm during G_2/M transition and its activity is strictly required in the first step of meiosis (Avazeri et al. 2000). DAG is unlikely to be the only second messenger generated by PtdIns(4,5)P_2 hydrolysis that is important for G_2/M transition. In fact, treatment with lithium of sea urchin embryos causes a reversible cell cycle arrest at metaphase, before nuclear envelope breakdown. This arrest is bypassed by photolysis of caged InsP$_3$, further indicating that nuclear PI signalling, possibly linked to calcium signals in and around the nucleus, is essential for proper timing of cell cycle transition (Becchetti and Whitaker 1997).

Assembly of the nuclear envelope is an essential step for the completion of mitosis and involves the binding of nuclear envelope vesicles to chromatin followed by membrane fusion. Recent experiments performed with sea urchin egg extracts highlighted a crucial role for a population of PI-enriched vesicles in nuclear envelope formation together with two distinct PtdIns-directed activities—a PtdIns-specific PLC and a PI3K activity—which have not yet been identified at the molecular level (Larijani et al. 2001).

4.6
Modulation of DNA Repair

The repair of double-strand breaks in DNA is crucial for genomic stability. Inability to repair double-strand breaks can cause loss of genetic information and chromosomal translocations leading to cell death. Two independent pathways guard cells from the dangers associated with double-strand breaks, homologous recombination, a process particularly used during S-phase, and nonhomologous end-joining (NHEJ), a general mechanism in place at all times during the cell cycle (Hopfner et al. 2002). Recently, both pathways have been shown to be regulated by inositol polyphosphates.

Homologous recombination is the main mechanism responsible for the repair of double-stranded DNA breaks (Friedberg 2003). PKC1 acts as a negative modulator of this process in *S. cerevisiae*, and its inactivation causes a dramatic hyper-recombination phenotype. Snyder and collaborators found that InsP$_6$ kinases (InsP$_6$ Ks) and their inositol pyrophosphate products, InsP$_7$ and InsP$_8$, are necessary for this process in mutant strains. Because InsP$_7$ and InsP$_8$ are required for hyper-recombination events requiring different sets of enzymes, it is likely that these inositol pyrophosphates have a general role in DNA recombination. A possible scenario foresees the transfer of their high-energy pyrophosphate groups to DNA or to proteins involved in this process. Inositol pyrophosphates might therefore provide a new means of protein phosphorylation, regulating distinct nuclear and cytoplasmic enzymes (Luo et al. 2002).

In mammals, DNA double-strand repair by NHEJ is dependent on several factors including DNA ligase IV, XRCC4 and the DNA-dependent protein kinase DNA-PK holoenzyme. The latter is formed by the heterodimeric complex Ku 70/80 and the catalytic subunit of DNA-PK (DNA-PKcs). $InsP_6$ has been shown to act as a key co-factor of this kinase and to stimulate DNA-PK-dependent double-strand DNA repair by NHEJ in vitro (Hanakahi et al. 2000; Hanakahi and West 2002). $InsP_6$ forms a stable complex with DNA-PK by binding to Ku 70/80 and causing a conformational change in the heterodimer (Hanakahi and West 2002). This transition might induce a change in the DNA-binding ability of DNA-PK and/or its interaction with the break site. Although the mechanism by which Ku 70/80 promotes end-to-end interaction is presently unclear, its ability to confine DNA movement is likely to be crucial in the precise pairing of the DNA ends. Moreover, $InsP_6$ might change the affinity of some of the DNA-PK interacting proteins, promoting the recruitment of essential factors at the repair site. Interestingly, the stimulatory activity of $InsP_6$ on NHEJ is restricted to higher eukaryotes, as the yeast orthologue of Ku 70/80 fails to bind $InsP_6$.

5
Concluding Remarks

In this review we have discussed only a fraction of the available evidence demonstrating the presence of PI and inositol polyphosphates in the nucleus, along with the enzymes involved in their metabolism. The emerging picture points towards a common mechanism whereby these inositol-containing molecules act as regulators of multi-subunit nuclear complexes involved in essential nuclear pathways. In particular, $PtdIns(4,5)P_2$ could function as a direct modulator of nuclear machineries by coupling them to the actin treadmill. The equilibrium between monomeric and polymeric forms of actin may regulate their sub-nuclear localisation and/or their activity, as directly shown for serum response factor-dependent gene transcription (Sotiropoulos et al. 1999). By acting as substrates of nuclear PLCs, PtdIns(4,5)P_2 could also provide localised sub-pools of InsP$_3$ for the generation of inositol polyphosphates, which act as essential co-factors for several nuclear processes, ranging from transcriptional control to RNA export and DNA repair. These compounds might therefore act as high turnover switches for these molecular machines, their activation dependent on the phosphorylation state of the inositol ring and restricted temporally and spatially based on their local synthesis and degradation. It will be interesting in the future to gain a global view of the specific sub-nuclear localisation of all these players, so as to further understand the functional relationships between them. In particular, we are unaware of any evidence for the existence of PtdIns(3,4)P_2 or PtdIns (3,5)P_2 in nuclei. Perhaps in the coming years, all members of the PI family will be shown to occur in nuclei as they do in the cytosol.

Acknowledgements. We sincerely apologise to all the colleagues whose work has been omitted due to space limitations. We thank Drs. J. Carroll, R. Irvine and S.L. Osborne for critical reading of the manuscript. This work was supported by Cancer Research UK.

References

Avazeri N, Courtot AM, Pesty A, Duquenne C, Lefevre B (2000) Cytoplasmic and nuclear phospholipase C-beta 1 relocation: role in resumption of meiosis in the mouse oocyte. Mol Biol Cell 11:4369–4380

Bacqueville D, Deleris P, Mendre C, Pieraggi MT, Chap H, Guillon G, Perret B, Breton-Douillon M (2001) Characterization of a G protein-activated phosphoinositide 3-kinase in vascular smooth muscle cell nuclei. J Biol Chem 276:22170-22176

Becchetti A, Whitaker M (1997) Lithium blocks cell cycle transitions in the first cell cycles of sea urchin embryos, an effect rescued by myo-inositol. Development 124:1099–1107

Berridge MJ (1993) Cell signalling. A tale of two messengers. Nature 365:388–389

Bertagnolo V, Marchisio M, Volinia S, Caramelli E, Capitani S (1998) Nuclear association of tyrosine-phosphorylated Vav to phospholipase C-γ1 and phosphoinositide 3-kinase during granulocytic differentiation of HL-60 cells. FEBS Lett 441:480–484

Borgatti P, Martelli AM, Bellacosa A, Casto R, Massari L, Capitani S, Neri LM (2000) Translocation of Akt/PKB to the nucleus of osteoblast-like MC3T3-E1 cells exposed to proliferative growth factors. FEBS Lett 477:27–32

Boronenkov IV, Loijens JC, Umeda M, Anderson RA (1998) Phosphoinositide signaling pathways in nuclei are associated with nuclear speckles containing pre-mRNA processing factors. Mol Biol Cell 9:3547–3560

Boudonck K, Dolan L, Shaw PJ (1999) The movement of coiled bodies visualized in living plant cells by the green fluorescent protein. Mol Biol Cell 10:2297–2307

Brazil DP, Hemmings BA (2001) Ten years of protein kinase B signalling: a hard Akt to follow. Trends Biochem Sci 26:657–664

Bunney TD, Watkins PA, Beven AF, Shaw PJ, Hernandez LE, Lomonossoff GP, Shanks M, Peart J, Drobak BK (2000) Association of phosphatidylinositol 3-kinase with nuclear transcription sites in higher plants. Plant Cell 12:1679–1688

Cairns BR, Erdjument-Bromage H, Tempst P, Winston F, Kornberg RD (1998) Two actin-related proteins are shared functional components of the chromatin-remodeling complexes RSC and SWI/SNF. Mol Cell 2:639–651

Calcerrada MC, Miguel BG, Martin L, Catalan RE, Martinez AM (2002) Involvement of phosphatidylinositol 3-kinase in nuclear translocation of protein kinase C zeta induced by C2-ceramide in rat hepatocytes. FEBS Lett 514:361–365

Chang SC, Miller AL, Feng Y, Wente SR, Majerus PW (2002) The human homolog of the rat inositol phosphate multikinase is an inositol 1,3,4,6-tetrakisphosphate 5-kinase. J Biol Chem 277:43836–43843

Ciruela A, Hinchliffe KA, Divecha N, Irvine RF (2000) Nuclear targeting of the beta isoform of type II phosphatidylinositol phosphate kinase (phosphatidylinositol 5-phosphate 4-kinase) by its alpha-helix 7. Biochem J 346:587–591

Clarke JH, Letcher AJ, D'Santos C S, Halstead JR, Irvine RF, Divecha N (2001) Inositol lipids are regulated during cell cycle progression in the nuclei of murine erythroleukaemia cells. Biochem J 357:905–910

Cocco L, Gilmour RS, Ognibene A, Letcher AJ, Manzoli FA, Irvine RF (1987) Synthesis of polyphosphoinositides in nuclei of Friend cells. Evidence for polyphosphoinositide metabolism inside the nucleus which changes with cell differentiation. Biochem J 248:765–770

Cockcroft S (1998) Phosphatidylinositol transfer proteins: a requirement in signal transduction and vesicle traffic. Bioessays 20:423–432

Cockcroft S (2000) Biology of phosphoinositides. In DM Glover (ed) Frontiers of Molecular Biology. Oxford University Press, Oxford, pp 341

Cooke FT (2002) Phosphatidylinositol 3,5-bisphosphate: metabolism and function. Arch Biochem Biophys 407:143–151

Cullen PJ, Cozier GE, Banting G, Mellor H (2001) Modular phosphoinositide-binding domains—their role in signalling and membrane trafficking. Curr Biol 11:R882–893

Czech MP (2002) Dynamics of phosphoinositides in membrane retrieval and insertion. Annu Rev Physiol 65:33.1–33.25

D'Santos CS, Clarke JH, Divecha N (1998) Phospholipid signalling in the nucleus. Biochim Biophys Acta 1436:201–232

D'Santos CS, Clarke JH, Irvine RF, Divecha N (1999) Nuclei contain two differentially regulated pools of diacylglycerol. Curr Biol 9:437–440

De Vries KJ, Westerman J, Bastiaens PI, Jovin TM, Wirtz KW, Snoek GT (1996) Fluorescently labeled phosphatidylinositol transfer protein isoforms (alpha and beta), microinjected into fetal bovine heart endothelial cells, are targeted to distinct intracellular sites. Exp Cell Res 227:33–39

Didichenko SA, Thelen M (2001) Phosphatidylinositol 3-kinase c2alpha contains a nuclear localization sequence and associates with nuclear speckles. J Biol Chem 276:48135–48142

Divecha N, Banfic H, Irvine RF (1991) The polyphosphoinositide cycle exists in the nuclei of Swiss 3T3 cells under the control of a receptor (for IGF-I) in the plasma membrane, and stimulation of the cycle increases nuclear diacylglycerol and apparently induces translocation of protein kinase C to the nucleus. EMBO J 10:3207–3214

Divecha N, Banfic H, Irvine RF (1993) Inosities and the nucleus and inosities in the nucleus. Cell 74:405–407

Drobak BK, Heras B (2002) Nuclear phosphoinositides could bring FYVE alive. Trends Plant Sci 7:132–138

Felsenfeld G, Groudine M (2003) Controlling the double helix. Nature 421:448–453

Feng Y, Wente SR, Majerus PW (2001) Overexpression of the inositol phosphatase SopB in human 293 cells stimulates cellular chloride influx and inhibits nuclear mRNA export. Proc Natl Acad Sci USA 98:875–879

Fricker M, Hollinshead M, White N, Vaux D (1997) Interphase nuclei of many mammalian cell types contain deep, dynamic, tubular membrane-bound invaginations of the nuclear envelope. J Cell Biol 136:531–544

Friedberg EC (2003) DNA damage and repair. Nature 421:436–440

Fukami K, Sawada N, Endo T, Takenawa T (1996) Identification of a phosphatidylinositol 4,5-bisphosphate-binding site in chicken skeletal muscle alpha-actinin. J Biol Chem 271:2646–2650

Fukami K, Furuhashi K, Inagaki M, Endo T, Hatano S, Takenawa T (1992) Requirement of phosphatidylinositol 4,5-bisphosphate for alpha-actinin function. Nature 359:150–152

Gillooly DJ, Morrow IC, Lindsay M, Gould R, Bryant NJ, Gaullier JM, Parton RG, Stenmark H (2000) Localization of phosphatidylinositol 3-phosphate in yeast and mammalian cells. EMBO J 19:4577–4588

Gimm O, Perren A, Weng LP, Marsh DJ, Yeh JJ, Ziebold U, Gil E, Hinze R, Delbridge L, Lees JA, Mutter GL, Robinson BG, Komminoth P, Dralle H, Eng C (2000) Differential nuclear and cytoplasmic expression of PTEN in normal thyroid tissue, and benign and malignant epithelial thyroid tumors. Am J Pathol 156:1693–1700

Gross SD, Loijens JC, Anderson RA (1999) The casein kinase Iα isoform is both physically positioned and functionally competent to regulate multiple events of mRNA metabolism. J Cell Sci 112:2647–2656

Guo K, Nichol R, Skehel P, Dormann D, Weijer CJ, Williams JG, Pears C (2001) A *Dictyostelium* nuclear phosphatidylinositol phosphate kinase required for developmental gene expression. EMBO J 20:6017–6027

Hanakahi LA, West SC (2002) Specific interaction of IP$_6$ with human Ku70/80, the DNA-binding subunit of DNA-PK. EMBO J 21:2038–2044

Hanakahi LA, Bartlet-Jones M, Chappell C, Pappin D, West SC (2000) Binding of inositol phosphate to DNA-PK and stimulation of double-strand break repair. Cell 102:721–729

Hinchliffe KA, Giudici ML, Letcher AJ, Irvine RF (2002) Type IIα phosphatidylinositol phosphate kinase associates with the plasma membrane via interaction with type I isoforms. Biochem J 363:563–570

Hocevar BA, Burns DJ, Fields AP (1993) Identification of protein kinase C (PKC) phosphorylation sites on human lamin B. Potential role of PKC in nuclear lamina structural dynamics. J Biol Chem 268:7545–7552

Hopfner KP, Putnam CD, Tainer JA (2002) DNA double-strand break repair from head to tail. Curr Opin Struct Biol 12:115–122

Horn PJ, Peterson CL (2002) Molecular biology. Chromatin higher order folding-wrapping up transcription. Science 297:1824–1827

Iborra FJ, Jackson DA, Cook PR (2001) Coupled transcription and translation within nuclei of mammalian cells. Science 293:1139–1142

Irvine RF (2002) Nuclear lipid signaling. Sci STKE 2002:RE13

Irvine RF, Schell MJ (2001) Back in the water: the return of the inositol phosphates. Nat Rev Mol Cell Biol 2:327–338

Kuriki H, Tamiya-Koizumi K, Asano M, Yoshida S, Kojima K, Nimura Y (1992) Existence of phosphoinositide-specific phospholipase C in rat liver nuclei and its change during liver regeneration. J Biochem (Tokyo) 111:283–286

Lachyankar MB, Sultana N, Schonhoff CM, Mitra P, Poluha W, Lambert S, Quesenberry PJ, Litofsky NS, Recht LD, Nabi R, Miller SJ, Ohta S, Neel BG, Ross AH (2000) A role for nuclear PTEN in neuronal differentiation. J Neurosci 20:1404–1413

Lallena MJ, Correas I (1997) Transcription-dependent redistribution of nuclear protein 4.1 to SC35-enriched nuclear domains. J Cell Sci 110:239–247

Larijani B, Barona TM, Poccia DL (2001) Role for phosphatidylinositol in nuclear envelope formation. Biochem J 356:495–501

Lee MS, Silver PA (1997) RNA movement between the nucleus and the cytoplasm. Curr Opin Genet Dev 7:212–219

Leser GP, Fakan S, Martin TE (1989) Ultrastructural distribution of ribonucleoprotein complexes during mitosis. snRNP antigens are contained in mitotic granule clusters. Eur J Cell Biol 50:376–389

Lewis JD, Tollervey D (2000) Like attracts like: getting RNA processing together in the nucleus. Science 288:1385–1389

Liu N, Fukami K, Yu H, Takenawa T (1996) A new phospholipase C δ 4 is induced at S-phase of the cell cycle and appears in the nucleus. J Biol Chem 271:355–360

Lu PJ, Hsu AL, Wang DS, Yan HY, Yin HL, Chen CS (1998) Phosphoinositide 3-kinase in rat liver nuclei. Biochemistry 37:5738–5745

Luger K, Mader AW, Richmond RK, Sargent DF, Richmond TJ (1997) Crystal structure of the nucleosome core particle at 2.8 Å resolution. Nature 389:251–260

Luo HR, Saiardi A, Yu H, Nagata E, Ye K, Snyder SH (2002) Inositol pyrophosphates are required for DNA hyperrecombination in protein kinase c1 mutant yeast. Biochemistry 41:2509–2515

Manzoli L, Billi AM, Rubbini S, Bavelloni A, Faenza I, Gilmour RS, Rhee SG, Cocco L (1997) Essential role for nuclear phospholipase C β1 in insulin-like growth factor I-induced mitogenesis. Cancer Res 57:2137–2139

Manzoli L, Billi AM, Faenza I, Matteucci A, Martelli AM, Peruzzi D, Falconi M, Rhee SG, Gilmour RS, Cocco L (1999) Nuclear phospholipase C: a novel aspect of phosphoinositide signalling. Anticancer Res 19:3753–3756

Maraldi NM, Zini N, Squarzoni S, Del Coco R, Sabatelli P, Manzoli FA (1992) Intranuclear localization of phospholipids by ultrastructural cytochemistry. J Histochem Cytochem 40:1383–1392

Martelli AM, Sang N, Borgatti P, Capitani S, Neri LM (1999) Multiple biological responses activated by nuclear protein kinase C. J Cell Biochem 74:499–521

Martelli AM, Gilmour RS, Bertagnolo V, Neri LM, Manzoli L, Cocco L (1992) Nuclear localization and signalling activity of phosphoinositidase C β in Swiss 3T3 cells. Nature 358:242–245

Martelli AM, Tabellini G, Borgatti P, Bortul R, Capitani S, Neri LM (2003) Nuclear lipids: New functions for old molecules? J Cell Biochem 88:455–461

Martelli AM, Neri LM, Gilmour RS, Barker PJ, Huskisson NS, Manzoli FA, Cocco L (1991) Temporal changes in intracellular distribution of protein kinase C in Swiss 3T3 cells during mitogenic stimulation with insulin-like growth factor I and bombesin: translocation to the nucleus follows rapid changes in nuclear polyphosphoinositides. Biochem Biophys Res Commun 177:480–487

Mazzotti G, Zini N, Rizzi E, Rizzoli R, Galanzi A, Ognibene A, Santi S, Matteucci A, Martelli AM, Maraldi NM (1995) Immunocytochemical detection of phosphatidylinositol 4,5-bisphosphate localization sites within the nucleus. J Histochem Cytochem 43:181–191

Metjian A, Roll RL, Ma AD, Abrams CS (1999) Agonists cause nuclear translocation of phosphatidylinositol 3-kinase γ. A G$\beta\gamma$-dependent pathway that requires the p110γ amino terminus. J Biol Chem 274:27943–27947

Nakielny S, Dreyfuss G (1999) Transport of proteins and RNAs in and out of the nucleus. Cell 99:677–690

Neri LM, Borgatti P, Capitani S, Martelli AM (1998) Nuclear diacylglycerol produced by phosphoinositide-specific phospholipase C is responsible for nuclear translocation of protein kinase C-α. J Biol Chem 273:29738–29744

Neri LM, Bortul R, Borgatti P, Tabellini G, Baldini G, Capitani S, Martelli AM (2002) Proliferating or differentiating stimuli act on different lipid-dependent signaling pathways in nuclei of human leukemia cells. Mol Biol Cell 13:947–964

Odom AR, Stahlberg A, Wente SR, York JD (2000) A role for nuclear inositol 1,4,5-trisphosphate kinase in transcriptional control. Science 287:2026–2029

Osborne SL, Meunier FA, Schiavo G (2001a) Phosphoinositides as key regulators of synaptic function. Neuron 32:9–12

Osborne SL, Thomas CL, Gschmeissner S, Schiavo G (2001b) Nuclear PtdIns(4,5)P$_2$ assembles in a mitotically regulated particle involved in pre-mRNA splicing. J Cell Sci 114:2501–2511

Payrastre B, Nievers M, Boonstra J, Breton M, Verkleij AJ, Van Bergen en Henegouwen PM (1992) A differential location of phosphoinositide kinases, diacylglycerol kinase, and phospholipase C in the nuclear matrix. J Biol Chem 267:5078–5084

Pederson T (2000) Half a century of "the nuclear matrix". Mol Biol Cell 11:799–805

Phair RD, Misteli T (2000) High mobility of proteins in the mammalian cell nucleus. Nature 404:604–609

Platani M, Goldberg I, Lamond AI, Swedlow JR (2002) Cajal body dynamics and association with chromatin are ATP-dependent. Nat Cell Biol 4:502–508

Rando OJ, Zhao K, Crabtree GR (2000) Searching for a function for nuclear actin. Trends Cell Biol 10:92–97

Rando OJ, Zhao K, Janmey P, Crabtree GR (2002) Phosphatidylinositol-dependent actin filament binding by the SWI/SNF-like BAF chromatin remodeling complex. Proc Natl Acad Sci U S A 99:2824–2829

Saiardi A, Caffrey JJ, Snyder SH, Shears SB (2000) Inositol polyphosphate multikinase (ArgRIII) determines nuclear mRNA export in *Saccharomyces cerevisiae*. FEBS Lett 468:28–32

Saiardi A, Erdjument-Bromage H, Snowman AM, Tempst P, Snyder SH (1999) Synthesis of diphosphoinositol pentakisphosphate by a newly identified family of higher inositol polyphosphate kinases. Curr Biol 9:1323–1326

Shears SB (2000) Transcriptional regulation: a new dominion for inositol phosphate signaling? Bioessays 22:786–789

Shen X, Xiao H, Ranallo R, Wu WH, Wu C (2003) Modulation of ATP-dependent chromatin-remodeling complexes by inositol polyphosphates. Science 299:112–114

Sindic A, Aleksandrova A, Fields AP, Volinia S, Banfic H (2001) Presence and activation of nuclear phosphoinositide 3-kinase C2β during compensatory liver growth. J Biol Chem 276:17754–17761

Sotiropoulos A, Gineitis D, Copeland J, Treisman R (1999) Signal-regulated activation of serum response factor is mediated by changes in actin dynamics. Cell 98:159–169

Spector DL, Fu XD, Maniatis T (1991) Associations between distinct pre-mRNA splicing components and the cell nucleus. EMBO J 10:3467–3481

Steger DJ, Haswell ES, Miller AL, Wente SR, O'Shea EK (2003) Regulation of chromatin remodeling by inositol polyphosphates. Science 299:114–116

Stevenson-Paulik J, Odom AR, York JD (2002) Molecular and biochemical characterization of two plant inositol polyphosphate 6-/3-/5-kinases. J Biol Chem 277:42711–42718

Sun B, Murray NR, Fields AP (1997) A role for nuclear phosphatidylinositol-specific phospholipase C in the G2/M phase transition. J Biol Chem 272:26313–26317

Sylvia V, Curtin G, Norman J, Stec J, Busbee D (1988) Activation of a low specific activity form of DNA polymerase α by inositol-1,4-bisphosphate. Cell 54:651–658

Takenawa T, Itoh T (2001) Phosphoinositides, key molecules for regulation of actin cytoskeletal organization and membrane traffic from the plasma membrane. Biochim Biophys Acta 1533:190–206

Thomas CL, Steel J, Prestwich GD, Schiavo G (1999) Generation of phosphatidylinositol-specific antibodies and their characterization. Biochem Soc Trans 27:648–652

Toker A (2002) Phosphoinositides and signal transduction. Cell Mol Life Sci 59:761–779

Vanhaesebroeck B, Leevers SJ, Ahmadi K, Timms J, Katso R, Driscoll PC, Woscholski R, Parker PJ, Waterfield MD (2001) Synthesis and function of 3-phosphorylated inositol lipids. Annu Rev Biochem 70:535–602

Vann LR, Wooding FB, Irvine RF, Divecha N (1997) Metabolism and possible compartmentalization of inositol lipids in isolated rat-liver nuclei. Biochem J 327:569–576

Verbsky JW, Wilson MP, Kisseleva MV, Majerus PW, Wente SR (2002) The synthesis of inositol hexakisphosphate. Characterization of human inositol 1,3,4,5,6-pentakisphosphate 2-kinase. J Biol Chem 277:31857–31862

Visnjic D, Crljen V, Curic J, Batinic D, Volinia S, Banfic H (2002) The activation of nuclear phosphoinositide 3-kinase C2β in all-trans-retinoic acid-differentiated HL-60 cells. FEBS Lett 529:268–274

Watt SA, Kular G, Fleming IN, Downes CP, Lucocq JM (2002) Subcellular localization of phosphatidylinositol 4,5-bisphosphate using the pleckstrin homology domain of phospholipase C δ1. Biochem J 363:657–666

Xu A, Suh PG, Marmy-Conus N, Pearson RB, Seok OY, Cocco L, Gilmour RS (2001) Phosphorylation of nuclear phospholipase C β1 by extracellular signal-regulated kinase mediates the mitogenic action of insulin-like growth factor I. Mol Cell Biol 21:2981–2990

Ye K, Hurt KJ, Wu FY, Fang M, Luo HR, Hong JJ, Blackshaw S, Ferris CD, Snyder SH (2000) PIKE. A nuclear GTPase that enhances PI3kinase activity and is regulated by protein 4.1N. Cell 103:919–930

York JD, Majerus PW (1994) Nuclear phosphatidylinositols decrease during S-phase of the cell cycle in HeLa cells. J Biol Chem 269:7847–7850

York JD, Xiong JP, Spiegelberg B (1998) Nuclear inositol signaling: a structural and functional approach. Adv Enzyme Regul 38:365–374

York JD, Odom AR, Murphy R, Ives EB, Wente SR (1999) A phospholipase C-dependent inositol polyphosphate kinase pathway required for efficient messenger RNA export. Science 285:96–100

Yu H, Fukami K, Watanabe Y, Ozaki C, Takenawa T (1998) Phosphatidylinositol 4,5-bisphosphate reverses the inhibition of RNA transcription caused by histone H1. Eur J Biochem 251:281–287

Zhao K, Wang W, Rando OJ, Xue Y, Swiderek K, Kuo A, Crabtree GR (1998) Rapid and phosphoinositol-dependent binding of the SWI/SNF-like BAF complex to chromatin after T lymphocyte receptor signaling. Cell 95:625–636

Zini N, Ognibene A, Bavelloni A, Santi S, Sabatelli P, Baldini N, Scotlandi K, Serra M, Maraldi NM (1996) Cytoplasmic and nuclear localization sites of phosphatidylinositol 3-kinase in human osteosarcoma sensitive and multidrug-resistant Saos-2 cells. Histochem Cell Biol 106:457–464

Subject Index

A
Actin Cross-Linking Protein 137
actin cytoskeleton 118–123, 125–127, 132, 134, 139, 141–142, 146–147
Actin Filament Nucleation 134
Actin Filament-Capping and -Severing Protein 127
Actin Monomer-Binding Protein 130
actin polymerization 118, 123, 125–126, 130, 134, 142, 146–147
actin- and PI(4,5)P2-binding site 132
actin-bundling 139
α-Actinin 137–139, 141
α-adaptin 37, 39
μ-adaptin 37
ADF/Cofilin 132–134, 143, 147, 149
ADP ribosylation factor (ARF) 170
AP2 32, 35, 37–38, 40
AP180 31–32, 38, 40, 42
apoptosis 92, 106
ARAP3 63, 65, 72
ARF 63–65, 72
β-ARK 66, 76
ARNO 63, 65, 72
β-arrestin 35, 37
autoinhibited 135
autophagy 105, 107–108

B
bind PI(3,4,5)P3 146
bind PI(4,5)P2 129, 145
Binding to PI(4,5)P2 140
binds actin monomer 134
binds to -actinin 138
binds to acidic phospholipid 138
Boyden chamber 167
Brutons tyrosine kinase (Btk) 52

C
CALM 38, 41–42
Capping Protein 127–128, 147
CCV 33, 37, 43
Cdc42 120, 126, 134–135, 142–144
Chemical structure 121
Chemotaxis 68–69, 71–73
Chromatin Remodelling 189, 191–193
Clathrin 31–33, 35, 37–39, 167
compass 167–168
Cortactin 135, 137
Cortexillin 139
CRAC 70, 73

D
DAPP1 53, 55, 57, 60–61, 63
depolymerizing 132
DH domain 67, 143–144
diacylglycerol 178, 186
Dictyostelium 168–170, 172
DNA Synthesis 196–197
dynamin 52

E
EEA1 93, 95–96, 98–99
Effects of elevated PI(4,5)P2 level 124
Endocytosis 33, 35, 38
ENTH domain 31, 33, 38, 40, 42–43
ERM Protein 139–142

F
F-actin 119, 127, 140–141
Fc receptors 169
FERM domain 76
filament 118–119, 123, 125, 127–130, 132, 134–135, 137–139, 142–143, 146–147, 149

filamin   119, 138
focal adhesion   138, 141
focal adhesion site   141
FYVE   90–91, 93–102, 104–105, 108–109, 167, 171

G
G-actin   119
GAP   143
GEF   64–65, 67, 72, 143–144, 147, 149
Gelsolin Family Protein   128–129
Gene Expression   189, 191
General Feature   121
GMC protein   125, 144–146
Golgi   64, 74–75
GPCR   69
Grp1   55, 57, 61
GTPase activity   143

H
Hip1R   32, 38
Hrs   95–96, 100, 105

I
induce actin assembly   123
inositol polyphosphate   177, 179, 186, 189, 191–192, 194–195, 197–199
inositol pyrophosphate   198
$\beta 3$ integrins   167
interact with PI(4,5)P2   141

L
lipid raft   125
LY294002   170–171
lysosome   5, 20–21

M
macrophage   2, 10–11, 14–15
membrane raft   145
Membrane Traffic / membrane trafficking   90–91, 93, 98, 104, 109
minus-end (pointed end)   119
Mitotic Regulation   197
modulate lipid signaling event   129
mRNA export   181, 194–196
multivesicular body   91–92
Myosin X   146

N
NADPH oxidase   106, 170
N-formyl-Met-Leu-Phe peptide (fMLP)   167
neutrophil   11
nucleus   177, 179, 181–184, 188–189, 191, 193–197, 199

P
p40phox   170–171
p47phox   170–171
p67phox   170
PDK-1   63, 65
PH   166, 168, 171
PH domain   49, 52, 54–55, 57, 59–68, 70–71, 73–77, 123, 125–126, 135, 138, 140, 143–144, 146, 149
phagocytic cup   146
phagocytosis   1–4, 8–9, 11, 13–17, 165–166, 169–170, 172
phagosome   1–2, 4–5, 8–9, 11, 13–14, 16–21, 90–93, 97–99, 104, 107, 109
PhdA   70, 73
3-phosphatase   126
5-phosphatase   126
Phoshophoinositide 3-kinases (PI3K) / PI 3-kinase   63, 65, 69–71, 73, 89–92, 107–108, 122, 126, 129, 131, 138, 146, 149, 166
phosphatidylinositol   166
Phosphatidylinositol 3-phosphate   89–90
Phosphoinositide   1–2, 4, 6, 8–10, 17–18, 21, 90, 93, 101, 104, 107, 121–123, 127, 129–130, 133–135, 137, 142, 144, 147, 149
Phosphoinositides   121, 126, 139–140, 143
phospholipase C   52, 186, 195
phosphorylation   140–143, 145–146, 149
PI derivative   122
PI(3,4)P2   126, 130–131, 134, 145–146
PI(3,4,5)P3 / PtdIns(3,4,5)P3   118, 120, 122–123, 125–126, 128–131, 134, 138, 142, 144, 146–147, 149, 179, 187–189

PI(4)P 5-kinase 122–123, 125, 141, 147
PI(4,5)P2 / PtdIns(4,5)P2 31–33, 35, 37–38, 40, 42–43, 118, 120, 122–131, 133–135, 138–140, 142, 144–147, 149
PI(4,5)P2 activation 142
PI(4,5)P2 binding site 129, 138
PI(4,5)P2 disrupts the actin-profilin complex 130
PI(4,5)P2 inhibit 137, 139
PI(4,5)P2 prevent 134
PI(4,5)P2-binding interface 133
PI3 phosphatase PTEN 168
plus-end (barbed end) 119
pre-mRNA processing 180–182, 193–194
priming 165, 171
process involving cytoskeletal rearrangement 126
profilin 130–131, 142
protein kinase B (PKB) 52
pseudopodia 169–170
PtdIns(3)P 179, 183, 188
PTEN 71, 167–168
PX / PX domain 38, 90–91, 93, 101–109

R
Rab4 99
Rab5 92–93, 98–99
Rabenosyn-5 99
rac 120, 122, 126, 134, 142–144
Rho 120, 142–144, 147, 149, 169
rho family / Rho family GTPase 118, 120, 142, 147, 149

S
signaling pathway 120, 135, 140
spatial marker 125
spectrin 52, 74
stimulated by PI(4,5)P2 135
synaptojanin 32–34

T
talin 125, 141
TAPP1 54, 58, 60–61, 63
TGF- 98, 101
three-dimensional network 138–139
transcription 178–179, 181–182, 188–189, 191, 194–195, 199
transforming growth factor- 98
turnover 119, 130, 132
Twinfilin 134, 147

U
ubiquitin 100
Uncapping of filament plus-ends by PI(4,5)P2 128
upregulated by Ca2+ 128

V
vinculin 137, 141–142
vital for interacting with PI(4,5)P2 131

W
WASP family 134, 136, 149
WASP/N-WASP 134
wortmannin 170

Z
Zigmond chamber 168

Current Topics in Microbiology and Immunology

Volumes published since 1989 (and still available)

Vol. 238: **Coffman, Robert L.; Romagnani, Sergio (Eds.):** Redirection of Th1 and Th2 Responses. 1999. 6 figs. IX, 148 pp. ISBN 3-540-65048-2

Vol. 239: **Vogt, Peter K.; Jackson, Andrew O. (Eds.):** Satellites and Defective Viral RNAs. 1999. 39 figs. XVI, 179 pp. ISBN 3-540-65049-0

Vol. 240: **Hammond, John; McGarvey, Peter; Yusibov, Vidadi (Eds.):** Plant Biotechnology. 1999. 12 figs. XII, 196 pp. ISBN 3-540-65104-7

Vol. 241: **Westblom, Tore U.; Czinn, Steven J.; Nedrud, John G. (Eds.):** Gastroduodenal Disease and Helicobacter pylori. 1999. 35 figs. XI, 313 pp. ISBN 3-540-65084-9

Vol. 242: **Hagedorn, Curt H.; Rice, Charles M. (Eds.):** The Hepatitis C Viruses. 2000. 47 figs. IX, 379 pp. ISBN 3-540-65358-9

Vol. 243: **Famulok, Michael; Winnacker, Ernst-L.; Wong, Chi-Huey (Eds.):** Combinatorial Chemistry in Biology. 1999. 48 figs. IX, 189 pp. ISBN 3-540-65704-5

Vol. 244: **Daëron, Marc; Vivier, Eric (Eds.):** Immunoreceptor Tyrosine-Based Inhibition Motifs. 1999. 20 figs. VIII, 179 pp. ISBN 3-540-65789-4

Vol. 245/I: **Justement, Louis B.; Siminovitch, Katherine A. (Eds.):** Signal Transduction and the Coordination of B Lymphocyte Development and Function I. 2000. 22 figs. XVI, 274 pp. ISBN 3-540-66002-X

Vol. 245/II: **Justement, Louis B.; Siminovitch, Katherine A. (Eds.):** Signal Transduction on the Coordination of B Lymphocyte Development and Function II. 2000. 13 figs. XV, 172 pp. ISBN 3-540-66003-8

Vol. 246: **Melchers, Fritz; Potter, Michael (Eds.):** Mechanisms of B Cell Neoplasia 1998. 1999. 111 figs. XXIX, 415 pp. ISBN 3-540-65759-2

Vol. 247: **Wagner, Hermann (Ed.):** Immunobiology of Bacterial CpG-DNA. 2000. 34 figs. IX, 246 pp. ISBN 3-540-66400-9

Vol. 248: **du Pasquier, Louis; Litman, Gary W. (Eds.):** Origin and Evolution of the Vertebrate Immune System. 2000. 81 figs. IX, 324 pp. ISBN 3-540-66414-9

Vol. 249: **Jones, Peter A.; Vogt, Peter K. (Eds.):** DNA Methylation and Cancer. 2000. 16 figs. IX, 169 pp. ISBN 3-540-66608-7

Vol. 250: **Aktories, Klaus; Wilkins, Tracy, D. (Eds.):** Clostridium difficile. 2000. 20 figs. IX, 143 pp. ISBN 3-540-67291-5

Vol. 251: **Melchers, Fritz (Ed.):** Lymphoid Organogenesis. 2000. 62 figs. XII, 215 pp. ISBN 3-540-67569-8

Vol. 252: **Potter, Michael; Melchers, Fritz (Eds.):** B1 Lymphocytes in B Cell Neoplasia. 2000. XIII, 326 pp. ISBN 3-540-67567-1

Vol. 253: **Gosztonyi, Georg (Ed.):** The Mechanisms of Neuronal Damage in Virus Infections of the Nervous System. 2001. approx. XVI, 270 pp. ISBN 3-540-67617-1

Vol. 254: **Privalsky, Martin L. (Ed.):** Transcriptional Corepressors. 2001. 25 figs. XIV, 190 pp. ISBN 3-540-67569-8

Vol. 255: **Hirai, Kanji (Ed.):** Marek's Disease. 2001. 22 figs. XII, 294 pp. ISBN 3-540-67798-4

Vol. 256: **Schmaljohn, Connie S.; Nichol, Stuart T. (Eds.):** Hantaviruses. 2001, 24 figs. XI, 196 pp. ISBN 3-540-41045-7

Vol. 257: **van der Goot, Gisou (Ed.):** Pore-Forming Toxins, 2001. 19 figs. IX, 166 pp. ISBN 3-540-41386-3

Vol. 258: **Takada, Kenzo (Ed.):** Epstein-Barr Virus and Human Cancer. 2001. 38 figs. IX, 233 pp. ISBN 3-540-41506-8

Vol. 259: **Hauber, Joachim, Vogt, Peter K. (Eds.)**: Nuclear Export of Viral RNAs. 2001. 19 figs. IX, 131 pp. ISBN 3-540-41278-6

Vol. 260: **Burton, Didier R. (Ed.)**: Antibodies in Viral Infection. 2001. 51 figs. IX, 309 pp. ISBN 3-540-41611-0

Vol. 261: **Trono, Didier (Ed.)**: Lentiviral Vectors. 2002. 32 figs. X, 258 pp. ISBN 3-540-42190-4

Vol. 262: **Oldstone, Michael B.A. (Ed.)**: Arenaviruses I. 2002, 30 figs. XVIII, 197 pp. ISBN 3-540-42244-7

Vol. 263: **Oldstone, Michael B. A. (Ed.)**: Arenaviruses II. 2002, 49 figs. XVIII, 268 pp. ISBN 3-540-42705-8

Vol. 264/I: **Hacker, Jörg; Kaper, James B. (Eds.)**: Pathogenicity Islands and the Evolution of Microbes. 2002. 34 figs. XVIII, 232 pp. ISBN 3-540-42681-7

Vol. 264/II: **Hacker, Jörg; Kaper, James B. (Eds.)**: Pathogenicity Islands and the Evolution of Microbes. 2002. 24 figs. XVIII, 228 pp. ISBN 3-540-42682-5

Vol. 265: **Dietzschold, Bernhard; Richt, Jürgen A. (Eds.)**: Protective and Pathological Immune Responses in the CNS. 2002. 21 figs. X, 278 pp. ISBN 3-540-42668-X

Vol. 266: **Cooper, Koproski (Eds.)**: The Interface Between Innate and Acquired Immunity, 2002, 15 figs. XIV, 116 pp. ISBN 3-540-42894-1

Vol. 267: **Mackenzie, John S.; Barrett, Alan D. T.; Deubel, Vincent (Eds.)**: Japanese Encephalitis and West Nile Viruses. 2002. 66 figs. X, 418 pp. ISBN 3-540-42783-X

Vol. 268: **Zwickl, Peter; Baumeister, Wolfgang (Eds.)**: The Proteasome-Ubiquitin Protein Degradation Pathway. 2002, 17 figs. X, 213 pp. ISBN 3-540-43096-2

Vol. 269: **Koszinowski, Ulrich H.; Hengel, Hartmut (Eds.)**: Viral Proteins Counteracting Host Defenses. 2002, 47 figs. XII, 325 pp. ISBN 3-540-43261-2

Vol. 270: **Beutler, Bruce; Wagner, Hermann (Eds.)**: Toll-Like Receptor Family Members and Their Ligands. 2002, 31 figs. X, 192 pp. ISBN 3-540-43560-3

Vol. 271: **Koehler, Theresa M. (Ed.)**: Anthrax. 2002, 14 figs. X, 169 pp. ISBN 3-540-43497-6

Vol. 272: **Doerfler, Walter; Böhm, Petra (Eds.)**: Adenoviruses: Model and Vectors in Virus-Host Interactions. Virion and Structure, Viral Replication, Host Cell Interactions. 2003, 63 figs., approx. 280 pp. ISBN 3-540-00154-9

Vol. 273: **Doerfler, Walter; Böhm, Petra (Eds.)**: Adenoviruses: Model and Vectors in Virus-Host Interactions. Immune System, Oncogenesis, Gene Therapy. 2004, 35 figs., approx. 280 pp. ISBN 3-540-06851-1

Vol. 274: **Workman, Jerry L. (Ed.)**: Protein Complexes that Modify Chromatin. 2003, 38 figs., XII, 296 pp. ISBN 3-540-44208-1

Vol. 275: **Fan, Hung (Ed.)**: Jaagsiekte Sheep Retrovirus and Lung Cancer. 2003, 63 figs., XII, 252 pp. ISBN 3-540-44096-3

Vol. 276: **Steinkasserer, Alexander (Ed.)**: Dendritic Cells and Virus Infection. 2003, 24 figs., X, 296 pp. ISBN 3-540-44290-1

Vol. 277: **Rethwilm, Axel (Ed.)**: Foamy Viruses. 2003, 40 figs., X, 214 pp. ISBN 3-540-44388-6

Vol. 278: **Salomon, Daniel R.; Wilson, Carolyn (Eds.)**: Xenotransplantation. 2003, 22 figs., IX, 254 pp. ISBN 3-540-00210-3

Vol. 279: **Thomas, George; Sabatini, David; Hall, Michael N. (Eds.)**: TOR. 2004, 49 figs., X, 364 pp. ISBN 3-540-00534-X

Vol. 280: **Heber-Katz, Ellen (Ed.)**: Regeneration: Stem Cells and Beyond. 2004, 42 figs., XII, 194 pp. ISBN 3-540-02238-4

Vol. 281: **Young, John A. T. (Ed.)**: Cellular Factors Involved in Early Steps of Retroviral Replication. 2003, 21 figs., X, 240 pp. ISBN 3-540-00844-6

Printing: Saladruck Berlin
Binding: Stürtz AG, Würzburg